Michael J. Oard

THE MISSOULA FLOOD CONTROVERSY

AND THE GENESIS FLOOD

Cover Design by Mark Wolfe

ISBN 0-940384-32-9

Creation Research Society Monograph Series: Number 13

Creation Research Society Books
6801 N. Highway 89
Chino Valley, AZ 86323-9186

Printed in the United States of America

Table of Contents

Dedication

This book is dedicated to Will Tebbs (1939-2002), a creation scientist from Portland, Oregon. Will was an avid student, as well as lecturer, on the geology of the Columbia River Gorge and the effects of the Lake Missoula Flood. He led many popular field trips to the Columbia Gorge area, including a few that I attended. Below is a picture of Will next to one of the largest granite erratic boulders (now in pieces) in the Columbia River Gorge. It is found several hundred feet above the river on the Washington side.

Will was born in Goldendale, Washington, just north of the Gorge and raised in White Salmon on the Washington side of the Columbia Gorge. He received a B.S. degree in science education from Central Washington College, a M.S. in mathematics from the University of New Mexico, and another M.S. in physical science at the University of Oregon. He taught high school and college math, as well as astronomy, for many years.

Will was one of the founders of the Portland-based Design Science Association (DSA). He served as president of DSA from March 1994 to January 1996. He served on the board of the association from 1994 to 2002. He was the association historian and gave many creationist presentations at DSA meetings and at churches. I will miss him and his enthusiasm for creationist geology and astronomy.

Acknowledgments

I am indebted to a number of people over the years who helped me on the manuscript, provided photos, or accompanied me in the field. I am especially grateful to members of the Design Science Association of Portland, Oregon, who accompanied me on many trips to the Channeled Scabland, the Columbia Gorge, and even canoeing down the Willamette River to observe the rhythmites. These include Will Tebbs, Dr. Keith Swensen, Dennis Bokovoy, Dr. Harold Coffin, John Hergenrather, and Steve Sparkowich. The book is dedicated to Will Tebbs who accompanied the author on several field trips. I also thank Jeff Aylor, Ed Nafziger, Jeff Norberg, Rick Thompson, and Michael Shaver for the time they spent showing me various deposits from glacial Lake Missoula and the Lake Missoula flood. I thank Tim Skertich for showing me the Susquehanna water gap in Pennsylvania. I appreciate Drs Carl Wieland and Tas Walker of AIG, Australia, for commenting on an early draft. I thank all the reviewers in the Creation Research Society, especially Drs George Howe and Emmett Williams. Special thanks goes to Dr. Harold Coffin, Dennis Bokovoy, and John Hergenrather for also reviewing an earlier draft. I appreciate all the work my wife, Beverly, has spent reworking my English. I thank Paul Kollas, Andrew Snelling, and John Morris for the use of illustrations and photos. I appreciate the time Peter Klevberg took in drawing many of the diagrams illustrating my ideas of the Recessional Stage of the Flood. I appreciate Dr Harold Coffin's permission to use his Beartooth fault illustration. I thank Dan Lethia and Daniel Lewis of AIG, U.S., for the artwork on a number of illustrations. Michael Shaver is to be thanked for providing the cover picture and two other pictures of Potholes Coulee, as well as guiding me to that location. Lastly, Mark Wolfe, my son-in-law, deserves special credit for drawing and reworking many of the illustrations and laying out the book for printing.

Preface

Most people have never heard of the glacial Lake Missoula flood or the scientific controversy surrounding it. Early in the1920's, J Harland Bretz (his first name is just the letter J) published evidence for a gigantic flood hundreds of feet (about 100 meters) deep, a hundred miles (160 kilometers) wide, ripping through the open expanse of eastern Washington and channeling through the Columbia Gorge. Bretz himself could hardly believe such a cataclysm was even possible, nevertheless, he reluctantly catalogued the mounting evidence he found. Bretz's work is an outstanding example of investigative science, yet he was met with scientific and personal vehemence; his theory was considered an "outrageous hypothesis." The geological establishment came out *en masse* to stamp out his "heresy" as they called it. They went to extreme and ridiculous lengths to explain away the huge mass of field data Bretz had accumulated. A few changed their minds when they actually saw the evidence for themselves, but most went their graves believing Bretz was undermining science itself.

What was the basis for Bretz's crazy hypothesis? Why were practically all the geologists of the day so outraged? Why did they go to such lengths to discredit Bretz and erect their own "outrageous hypotheses"? Does such an incident have any meaning for science in general or historical geology in particular? Does such an affair unveil attitudes that continue to this day? This book will address these questions and briefly document the massive evidence that inspired Bretz to postulate what is now called the Lake Missoula flood. It will describe the angst over connecting the huge glacial Lake Missoula with the evidence for the gigantic flood in eastern Washington. After 40 years of contention and rejection, Bretz's flood is now accepted by mainstream geologists as having occurred during the peak of the ice age. Once the Lake Missoula flood was accepted, other catastrophic ice age floods were discovered or postulated in other areas of the world. Today, the controversy has shifted from whether there was one or dozens of Lake Missoula floods. Most scientists have come to believe there were many floods, but there is convincing evidence that there was only one major flood. Bretz's hypothesis was rejected because the Lake Missoula flood was considered too catastrophic; it was too close to the biblical Flood. This was the only reason. Geologists of Bretz's day thought they had dismissed catastrophes of any sort one hundred years earlier. Could the tale of the Lake Missoula flood indicate that the geologists were inspired by prejudice against the Genesis Flood? Could geologists today, because of this same tradition, likewise be overlooking geological evidence that there was a global Flood? Could this same blindness lead them to postulate long periods of time? I will explore these questions in this book while telling the story of the Lake Missoula flood.

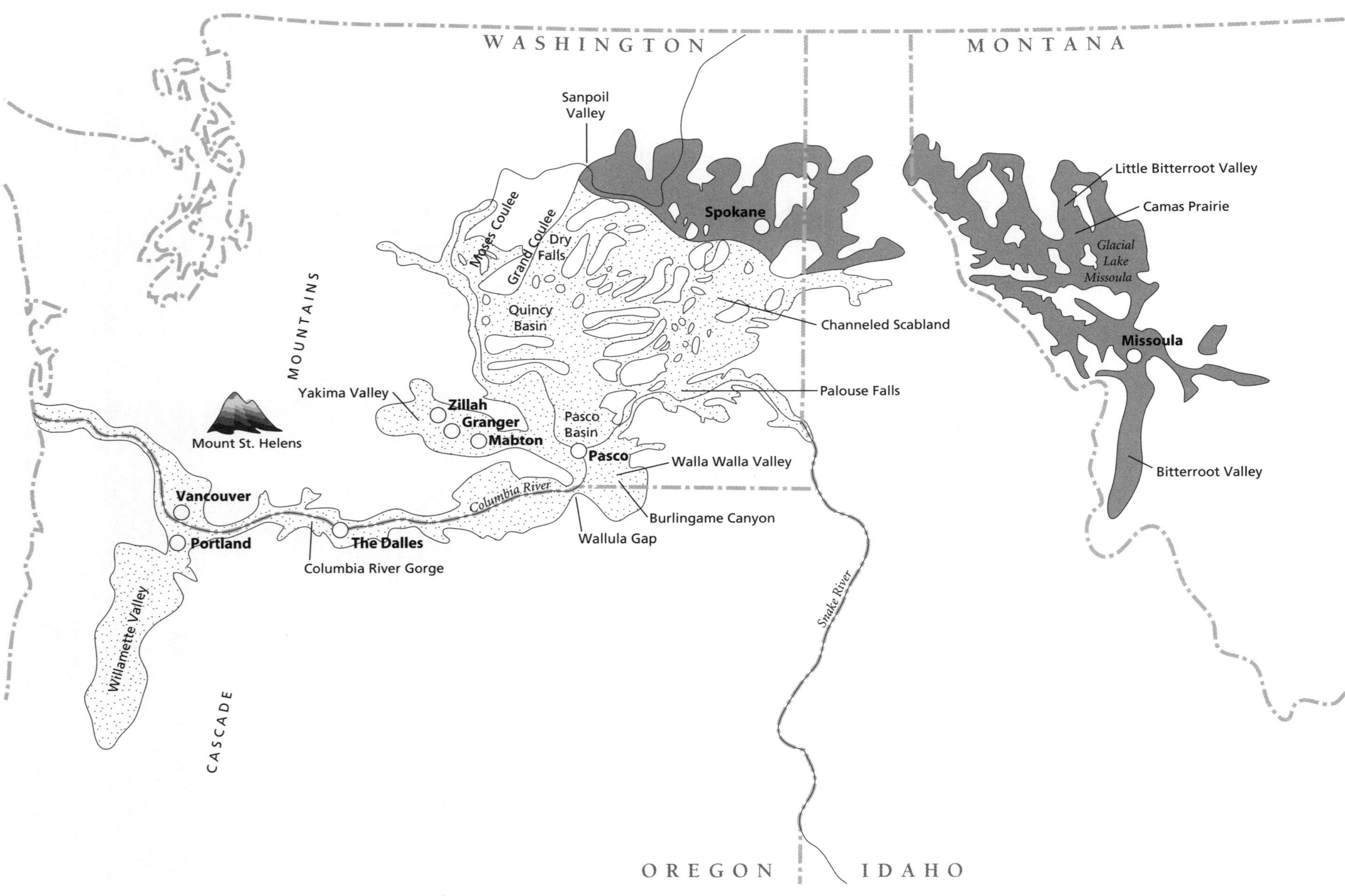

Figure 1.1 Map of the Pacific Northwest showing glacial Lakes Missoula and Columbia along with the Channeled Scabland of eastern Washington.

Chapter 1

The Flood of Controversy

Today it is hard to imagine the storm of controversy that was ignited when the geological community heard J Harland Bretz's hypothesis of a gigantic flood that roared through eastern Washington. To the scientists of the day it sounded far too much like the Biblical Flood, a Flood that scientists were convinced never occurred. It was not until the evidence for the Lake Missoula flood became overwhelming and irrefutable that it was finally accepted–40 years later.

BRETZ'S "OUTRAGEOUS HYPOTHESIS"

In the 1920s J Harland Bretz proposed a monstrous flood in the Pacific Northwest of the United States. He deduced that each coulee in eastern Washington was once a channel that was formed only if the *entire* coulee had been filled with water. A coulee is a generally dry trench-like valley with vertical walls and a flat floor. Since the coulees intersect, and the major coulee tracts head at the same elevation, he concluded that water had to be running through all of them *at the same time*. This area of eastern Washington is called the Channeled Scabland because of all the intersecting coulees. The width of the flood was about 100 miles (160 kilometers). Even more alarming to the scientific community, he calculated the flow depth at hundreds of feet (about 100 meters). Bretz first unveiled his hypothesis in 1923 in the *Bulletin of the Geological Society of America* (Bretz, 1923a) and the *Journal of Geology* (Bretz, 1923b). He suggested the source of the floodwaters was the rapidly melting Cordilleran Ice Sheet to the north, which he mistakenly believed had spread south of Spokane. (Later it was learned that the ice sheet remained north of Spokane.) The topographical evidence caused him to conclude the flood originated from somewhere around Spokane, so Bretz called it the Spokane flood. Unconvinced geologists had another name for it; they called it the "Bretz flood." To them the flood was truly an "outrageous hypothesis." It was only later, after it was accepted, that it was renamed the Lake Missoula flood after the source of the water. Figure 1.1 is a map of the Pacific Northwest showing the features of the scoured rock of eastern Washington.

Because the theory seemed so ridiculous to the scientific community, numerous well-known geologists became outraged and took action. In 1927, Bretz was invited for a showdown at the Geological Society in Washington, D.C., where they verbally ambushed him (Baker, 1978a, pp. 7-9). An array of bewildering hypotheses and objections were flung at Bretz, but he held his ground.

During all this early contention, Bretz quietly continued his fieldwork in eastern Washington. He compiled a massive case for the Spokane flood (Bretz, 1928c). During this time he was open to alternative hypotheses, since he was just as perplexed by the scale of the event as his critics. He carefully analyzed all the many hypotheses invented to explain away the observations. He found the evidence easily refuted these hypotheses (Bretz, 1928a, b). After nearly ten years, Bretz completed his research and submitted a monograph conclud-

Figure 1.2 Upper Grand Coulee, north central Washington (view north from near southern Banks Lake).

Figure 1.3 Lower Grand Coulee and the Coulee Monocline (view south from Coulee City, Washington).

ing that the Upper Grand Coulee (Figure 1.2) and the upper parts of the Lower Grand Coulee (Figure 1.3) that encompasses Dry Falls were the result of gigantic receding waterfalls (Bretz, 1932).

Bretz's own words attest to the fact that he accepted the flood hypothesis with great trepidation and searched for an alternative hypothesis:

> I think I am as eager as anyone to find an explanation for the Channeled Scabland of the Columbia Plateau that will fit all the facts and will satisfy geologists. I have put forth the flood hypothesis only after much hesitation and only when accumulating data seemed to offer no alternative (Bretz, 1927, p. 468).

Bretz (1930a, p. 422) was not given to wild, sensational claims and seems to have been a traditional uniformitarian geologist:

> The writer, at least normally sensitive to adverse criticism, has no desire to invite attention simply by advocating extremely novel views. Back of the repeated assertions of the verity of the Spokane Flood lays a unique assemblage of erosional forms and glacial water deposits: an assemblage which can be resolved into a genetic scheme only if time be very short, volume very large, velocity very high, and erosion chiefly by plucking of the jointed basalt.

Uniformitarianism is the philosophy that all past geological processes work at about the same rate as we see today. It can be remembered by the simple phrase *the present is the key to the past*. An easy way of thinking about uniformitarianism is to think of the word *uniform*–uniform, present processes shaped the world we see. Within this philosophy it would take millions of years for landforms, such as seen in eastern Washington, to be shaped by erosion. Because of his thorough research and his slowness in postulating the Spokane flood hypothesis, Bretz was better able to counter the barrage of criticism that fell upon him. In the early 1930s, Bretz changed his mind about the flood emanating from the melting of the Cordilleran Ice Sheet (Bretz, 1930b; Baker, 1978a, p. 9) and concluded instead that the bursting of glacial Lake Missoula caused the Spokane flood. This was a reasonable deduction that should not have generated any controversy, given the size of the lake and its upstream location. As will be discussed in Chapter 3, the evidence for glacial Lake Missoula had been known since the late 1800s. But, in the typical uniformitarian lockstep of the era, Joseph Pardee declared the glacial lake had drained "slowly." There is proof, however, that Pardee was not fully convinced of his own published beliefs, but in truth, thought the ice dam really burst catastrophically and caused the Lake Missoula flood. This was even *before* Bretz published his research, but his superiors who were dead set against any type of catastrophe dissuaded him against suggesting a catastrophic flood:

> Indeed, there are indications that in 1922 Pardee possessed key evidence supporting Bretz's hypothesis, but that he was dissuaded from revealing this by his superior, W.C. Alden, and by other colleagues of the U.S. Geological Survey...It is also probable that in 1922, before Bretz's work on the topic, Pardee had independently realized the cataclysmic origin of the Channeled Scabland, but had been prevented from publishing the idea by his superiors. Until 1940 he had remained silent on the topic (Baker et al., 1987, p. 418).

This shows the strength of peer pressure and the desire to conform to *preconceived* geological assumptions in spite of the evidence.

Bretz, himself, very likely knew of the existence of glacial Lake Missoula by 1925, but for some inexplicable reason chose to ignore the significance of Pardee's evidence for such a monstrous lake dammed by ice. Even up until 1932, he thought the bursting of the pro-glacial lake was only a *possibility* (Waitt, 1994, pp. k2-k3; Baker, 1995). It could have been that Bretz was overly cautious at this point, having endured years of ridicule.

OUTRAGEOUS ALTERNATIVE HYPOTHESES

Figure 1.4 Columbia River Gorge of the Columbia River, a large water gap between Oregon and Washington. (View west from White Salmon, Washington).

Bretz's critics fell all over themselves trying to explain his provocative observations in the Channeled Scabland. They developed many alternative hypotheses. Looking back at these alternative hypotheses from the vantage point of history demonstrates how desperate geologists were to oppose any hint of catastrophism, the idea that large catastrophes shaped the surface of the earth. Convinced of uniformitarianism, they could only accept slow changes over immense periods of time. They were more comfortable with wacky alternatives that were contrary to field evidence rather than to give an inch towards catastrophism.

W.C. Alden, one of the chief opponents who believed it was impossible for the Cordilleran Ice Sheet to produce so much water under any condition, remarked that the Channeled Scabland was simply formed by collapsed lava tubes. The bars, discussed in the next chapter, were the result of repeated flooding from much smaller volumes of water.

O.E. Meinzer, the father of modern hydrology, postulated a commonly held view that the Channeled Scabland was simply caused by the shifting of the course of the mighty Columbia River, a river that was much larger during the ice age. He concluded that swollen ice age rivers could easily have cut Dry Falls and deposited the great Ephrata gravel fan in Quincy Basin. He countered what he called Bretz's "violent claim," that all four spillways of the Quincy Basin were filled with water at the same altitude and at the same time, by contending the spillways were cut at different times and at different altitudes. He claimed that later earth movements eventually brought them all to the same altitude.

E.T. McKnight suggested the scablands were formed by erosion along the edge of the ice sheet as it shifted locations. Bretz had earlier concluded that the hanging valleys in the *soft* sediments of the Koontz channels, 150 to 400 feet (46 to 120 meters) higher than the Columbia River near Hanford, Washington, could have formed only if the Columbia was flowing *above* the altitude of the highest overhang (see chapter 2). This meant that the flood was up to 400 feet (120 meters) deep! McKnight (1927), who actually took part in the topographic mapping of the area, countered Bretz by suggesting that the Columbia River deepened its channel in post-glacial time leaving the Koontz channels hanging. Bretz (1927) demonstrated that McKnight's hypothesis did not work, since McKnight had ignored the unconsolidated nature of the sediments.

James Gilluly, in typical uniformitarian lockstep, agreed with others and simply stated that the channels of eastern Washington were the result of long-continued erosion and deposition by swollen ice age rivers. He easily showed that Bretz's two suggested sources for the floodwater were inadequate. As a result, Gilluly's arguments were all the more persuasive.

G.R. Mansfield also claimed without evidence that the scabland channels were not occupied by water at the same time. He thought the scablands were better explained by persistent ponding and the overflowing of lakes at the edge of the shifting ice sheet.

As for the streamlined and scarped silt "islands" that overlie the basalt of eastern Washington within the channels, G.O. Smith proposed that the islands were a product of rainwash and erosion caused by wind rising up over the lower basalt cliffs.

H.G. Ferguson suggested that river ice jams may have ponded the water in channels high enough to overflow onto the silt tracts forming high-divide crossings that Bretz had pointed as proof of water depth. But, the divided crossings are several hundred feet above the scabland floor!

On and on the hypotheses were invented. Pardee, who was an expert in the local geology and should have known better, simply stated that the scabland tracts were the results of "unusual glaciation." Pardee obviously was trying to avoid the subject, probably secretly agreeing with Bretz, but was fearful for his career and reputation (Allen, Burns, and Sargent, 1986, p. 56).

Amazingly, most of the geologists who criticized Bretz's

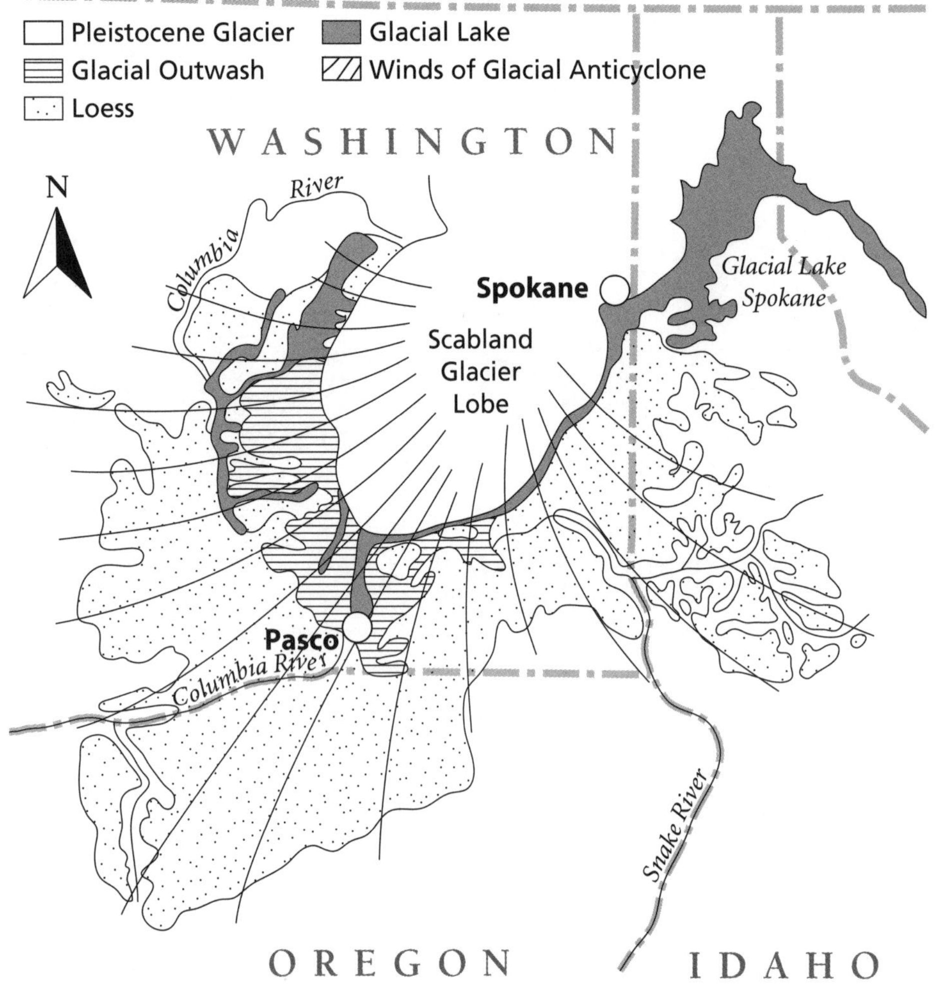

Figure 1.5 Hobbs' glacial lobe compared to the boundary of ice sheet in eastern Washington. (Redrawn from Hobbs, 1943, Figure 1 by Mark Wolfe).

Spokane flood hypothesis were *unfamiliar* with the field data of eastern Washington; they either had superficially examined certain sections in the field or *they had never been there*! Bretz was well able to counter all of these poorly supported criticisms (Bretz, 1927; 1928a, b). It is interesting that James Gilluly late in life finally visited the Channeled Scabland and observed the overwhelming evidence for such a flood. He was one of the few who eventually accepted the Spokane flood, remarking, "How could *anyone* have been so wrong?" (Baker, 1978a, p. 9)

THE BLINDNESS OF THE PARADIGM

Over the 40-year period that Bretz's hypothesis was rejected, several geologists did go out in the field to examine the evidence in depth. They developed rather sophisticated counter hypotheses but still failed to see the significance of the Channeled Scabland in spite of analyzing the Channeled Scabland firsthand. Their hypotheses were as bizarre as Bretz's original hypothesis appeared to be. They were blinded by the uniformitarian paradigm.

Ira Allison (1933), a geologist from Oregon State University, was familiar with the geology of the area and at least agreed with most of Bretz's field evidence. He certainly did not agree with Bretz's conclusions. Instead, he hypothesized that the unique features in eastern Washington were caused by a huge ice jam, aided by a landslide, in the Columbia River Gorge (Figure 1.4). In order to produce all the unique flooding features upstream in the Channeled Scabland, the ice jam had to grow to an incredible height of about 1100 feet (335 meters)! He further proposed that smaller ice jams upstream in eastern Washington produced some of the unique high elevation fluvial features, such as divide crossings. He did recognize problems associated with his hypothesis and its Herculean requirements, as he admitted, but was motivated solely by a desire to refute Bretz's flood theory (Allison, 1933, p. 677).

Edwin Hodge (1934), also a professor of geology at Oregon State, published a brief paper countering Bretz that involved mainly glacial processes. His hypothesis involved a complicated alternation of drainage changes with ice advances and retreats. According to Hodge, the channels were quarried by glacial erosion or by the diversion of meltwater around blocks of stagnant ice or ice jams. The theory was presented in outline form only, but he gave numerous seminars and oral papers attacking Bretz's Spokane flood. Baker (1981, p. 123) tells of an ugly confrontation between Bretz and Hodge at a scientific meeting in Pullman, Washington:

> Bretz had been invited to keynote the meeting with an address on the Spokane Flood. Hodge was so incensed that he wrote to the meeting chairman demanding that he be allowed to debate Bretz formally during the address. This request was refused, but, undaunted, Hodge then invented a subsidiary society with a concurrent meeting time.

This is an outrageous example of attempted censorship to suppress what was believed to be an unscientific hypothesis.

After a few weeks in eastern Washington, William Hobbs (1943), an eminent glacial geologist, reported the "new

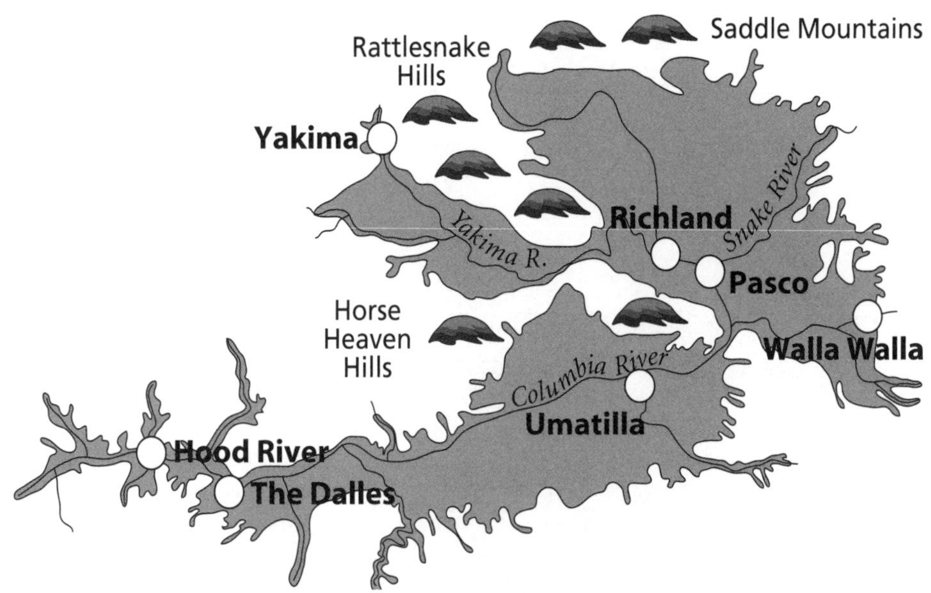

Figure 1.6 "Lake Lewis" in southeastern Washington. The lake was based on obvious high water marks at Wallula Gap suggesting upstream ponding. (Redrawn from Flint, 1938, Figure 5 by Mark Wolfe.)

discovery" of a lobe of the Cordilleran Ice Sheet that supposedly descended all the way down into extreme southeast Washington (Figure 1.5). Similar to Hodge's ideas, this "Scabland glacier lobe" formed all the scabland features. It is interesting how Hobbs describes his first "enlightenment." While listening to a lecture on the Spokane flood by Bretz, himself, he was "immediately struck" by the "...obvious signs of glacial intrusion that showed up on the 'lecturer's map.'" (Allen, Burns, and Sargent, 1986, p. 66). Victor Baker (1978a, p. 12) considers Hobbs' hypothesis full of *fundamental errors*:

> The Hobbs paper contains so many fundamental errors that one marvels at the absurd limits that were being stretched to find an alternative to catastrophic flooding as the cause of the Channeled Scabland. Hobbs...argued that the scabland was a product of glacial scour and that the Palouse loess [wind-blown silt] was deposited contemporaneous to this glaciation by anticyclonic winds off the ice that lay in the various "channels." He interpreted many scabland gravel deposits as moraine remnants modified by glacier-border drainage.

His opposition to the Spokane flood hypothesis and his allegiance to the doctrine of uniformitarianism blinded Hobbs.

The strongest challenge to Bretz came from Richard Foster Flint (1938), renowned ice age geologist from Yale University, who spent considerable time examining the relevant geological features in eastern Washington. Flint, with Ivy League authority, announced that glacial sediments were deposited in a large lake, "Lake Lewis," throughout eastern Washington (Figure 1.6). Flint believed that a huge landslide in the Columbia River Gorge that blocked the Columbia River caused this lake. This landslide had to pile debris to about 1100 feet (335 meters) deep above the Columbia River! He concluded that when the blockage on the Columbia River failed, it formed many of the unique features of the area by subsequent erosion. Flint cited a considerable amount of field data, resulting in a seemingly believable alternate interpretation to Bretz's hypothesis–all of it wrong! Flint also used the basalt scabland that was carved in the upper Snake River Valley, Idaho, as an example of what he thought was normal stream erosion (Baker, 1981, p. 126). As it later turns out, these scablands in Idaho are the product of another catastrophic flood, the Bonneville Flood (O'Connor, 1993), described in Chapter 6.

Even Allison (1941), who had a similar hypothesis as Flint, but with the Columbia Gorge blocked by ice instead of landslide debris, found many serious misinterpretations in Flint's hypothesis. Baker considers Flint's hypothesis one of the most ironic in the annals of geology, because Flint analyzed in detail the very deposits favorable to the giant flood hypothesis. Flint was able to elaborately fit the field data into his own uniformitarian hypothesis. Numerous giant gravel bars (Figure 1.7) cited by Bretz as caused by a gigantic flood were instead seen by Flint as "modified river terraces from leisurely streams with normal discharge." Such is the blinding power of uniformitarianism!

Flint wrote several books on the ice age. Even after Bretz was vindicated, Flint did not even mention the Channeled Scabland in a comprehensive textbook on the ice age published in 1957. In the third edition of this widely used treatise on the ice age, published in 1971, he provides a single, rather vague, sentence about the Channeled Scabland. Baker (1981, p. 167) states the irony of all this:

> The irony remained that this oft-cited textbook, noted for its thorough coverage, actually failed to mention the most spectacular event to occur during the Pleistocene [ice age] on our planet.

He goes on to state how Flint's hypothesis continued in the textbooks long after it was discredited and Bretz's hypothesis accepted:

> Moreover, despite its discrediting and Flint's own reluctance to mention the idea after 1956, [Flint's] scabland fill hypothesis continued to appear in texts through the 1960s (Baker, 1981, p. 167).

Figure 1.7 Crescent Gravel Bar along Columbia River, south of Wenatchee, Washington. Note ripple marks on upper surface, which are more clearly seen from the air.

Erroneous geological ideas die slowly.

One lesson to be learned from the critics of Bretz's outrageous hypothesis is that no matter how much field data is accumulated, the paradigm shapes what the scientists perceives, blinding him to all other possibilities.

BRETZ'S FLOOD BECOMES UNIFORMITARIAN

Even before Bretz's last in a series of publications came out in 1932, the Great Depression had hit and within eight years, World War II. Travel was restricted, so it was difficult for geologists to examine Bretz's field evidence. Opinion remained strongly against Bretz during this time. But cracks in this uniformitarian Jericho began to show up when J.T. Pardee finally published further evidence for the existence and drainage of glacial Lake Missoula (Pardee, 1940, 1942). Pardee documented the channeled wind gaps on ridges, giant gravel beds, and gravel ripple marks on Camas Prairie at relatively high elevations above the Clark Fork River and its tributaries. But still he was overly cautious and only weakly suggested that these deposits "might" have been caused by a rapid draining of glacial Lake Missoula. Geologists at that time seemed unable to recognize that, if a huge glacially ponded lake existed upstream, it could burst catastrophically and form the Channeled Scabland.

With the coming of a new generation of geologists and further fieldwork, the Lake Missoula flood was finally accepted in the 1950s and 1960s. Bretz, himself, returned to the field of eastern Washington in 1952 at age 70, accompanied by a skeptic of the Lake Missoula flood, H.T.U. Smith. The skeptic was soon convinced by Bretz, especially when the field evidence clearly established that the discontinuous giant mounds of gravel were indeed of flood-bar origin. By then, new evidence aided by aerial photography displayed the many giant ripple marks on the bar deposits (Bretz, Smith, and Neff, 1956; Bretz, 1969). The signs of a gigantic flood were obvious and multitudinous.

The Lake Missoula flood has now become widely accepted and quietly incorporated into the uniformitarian paradigm. As part of uniformitarianism, it became easier to visualize multiple Lake Missoula floods, since repeated events seem more uniformitarian. The question of the number of Lake Missoula floods now occupies modern debate. Before I address the subject of the number of floods, I will briefly describe the compelling evidence that led Bretz to his outrageous conclusion and the connection of the Spokane flood to the bursting of glacial Lake Missoula.

Chapter 2

The Incredible Channeled Scabland

The discovery of the Lake Missoula flood is one of the most outstanding investigative studies in the history of modern geology. Here is the story.

FOUR HUNDRED FOOT WATERFALLS – NOW DRY

After graduating from college during the early part of the 20th century, J Harland Bretz (1882-1981) began as a high school science teacher at Franklin High School in Seattle, Washington. While teaching for four years, he developed an interest in the glacial geology of the Puget Sound area. He spent much of his spare time studying these deposits, even writing papers for geological journals. He learned much of the geology of this area on his own in the field and from regular visits to the Department of Geology at the University of Washington.

Early in the 20th century the United States Geological Survey published topographic maps of Washington and Oregon. Bretz's curiosity was first aroused when he viewed the topographic map of the Quincy Basin of central Washington (see Figure 1.1 for place names) (Baker *et al.*, 1987, p. 417). Of particular interest to Bretz were two parallel canyons cut back into the basalt along the west central edge of Quincy Basin above the Columbia River (Figure 2.1). Bretz contemplated how these coulees could have formed. He reasoned that the most likely possibility was that two huge waterfalls separated by a narrow basalt ridge cut these coulees. An ancient river would have carved a gradual descent westward to the Columbia River. It looked as though a huge rush of water from Quincy Basin to the east had eroded the two notches in the basalt and receded eastward for about 2 miles (3 kilometers). Then the water mysteriously disappeared leaving behind the "dry" waterfalls. These two notches are known today as the Potholes Coulee or Cataract (Figure 2.2 and cover picture). Bretz also noticed two other notches in the ridge on the topographic maps. One is Crater Coulee on the northwest edge of Quincy Basin. The other is the Frenchman Springs Coulee cut into the southwest edge of the Quincy Basin (Figure 2.3). Frenchman Springs Coulee is similar to Potholes Coulee in that it also is a double coulee separated by a very thin basalt ridge. The size of these coulees is amazing. The walls of Potholes Coulee and Frenchman Springs Coulee are about 400 feet (120 meters) high and 600 feet (185 meters) above the Columbia River to the west.

Bretz also noticed another abandoned cataract in the middle of Grand Coulee, north of the Quincy Basin, now named Dry Falls (Figure 2.4 and 2.5). Dry Falls is about 4 miles (6 kilometers) wide and is also 400 feet (120 meters) high–when the water flowed it would have been many times larger than Niagara Falls.

Bretz found it interesting that Potholes Coulee, Frenchman Springs Coulee, and Dry Falls were composed of two horseshoe-shaped alcoves, separated by isolated plateau remnants, similar to Goat Island on Niagara Falls. Goat Island separates the American Falls from the Canadian Falls. At the base of each alcove at Dry Falls and the Potholes Cataract are lakes as deep as 115 feet (35 meters) that suggested to him that they might

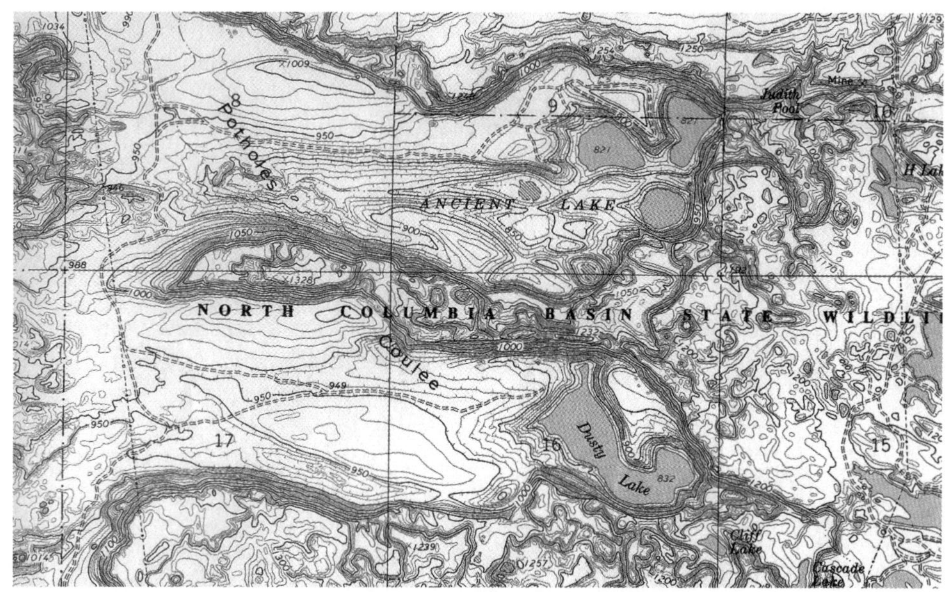

Figure 2.1 Topographic map of the Potholes Coulee, western Quincy Basin (Babcock Ridge, Washington, 7.5 minute quadrangle, 1966, courtesy of the United States Geological Survey).

be overdeepened plunge pools. A series of plunge-pool like lakes are observed "downstream" from the main plunge pools below the "falls." He also puzzled over the numerous large basalt boulders, some as large as houses, strewn along these excavated areas (Figure 2.6).

Figure 2.2 The northern coulee of the double Potholes Coulee on the west central edge of Quincy Basin, Washington (photo courtesy of Michael Shaver).

Bretz found these features mysterious because of the assumptions he was applying. He had been trained in the geological tradition of uniformitarianism like all geologists of his time, as well as today. Yet, the landforms he observed on the maps did not seem to fit this thinking. Bretz puzzled over the source of water needed to produce waterfalls of such gigantic proportions on a ridge in a near desert.

MISFIT BOULDERS

Bretz continued to ponder the meaning of the dry waterfalls when he left his teaching position and returned to college. He received his PhD degree with honors in geology from the University of Chicago. His thesis was based on his field notes of the glacial geology from the Puget Sound area. After graduating he became an assistant professor of geology at the University of Washington. During this time he investigated the Columbia River Gorge that runs through the Cascade Mountains between Washington and Oregon (see Figure 1.1). He noted even more puzzling geological features.

The Gorge is the most impressive water gap in the Pacific Northwest. A water gap is a valley that dissects a transverse barrier, like a mountain ridge, in which a river appears to have cut the path. However, it was not this unusual geological characteristic that caught Bretz's attention. It was the "misfit" boulders above the Columbia River that he noticed (see figure on dedication page). These misfit boulders are called erratic or exotic boulders because these types of rocks do not outcrop in the vicinity. Most of the boulders he observed were composed of granite. The nearest source for granite upstream from the Columbia River is about 200 miles (320 kilometers) away in northern Washington. Bretz became convinced that there was only one mechanism that could account for the erratic boulders. Sometime in the past there must have been a huge flood, one large enough to float *icebergs* that were capable of carrying huge boulders. The boulders were released after the iceberg became stranded and melted. Amazingly, these boulders lay hundreds of feet (over 100 meters) above the Columbia River!

Figure 2.3 Frenchman Springs Coulee on the southwest edge of Quincy Basin, Washington.

We observe misfit boulders throughout eastern Washington, the Columbia Gorge, and into the Willamette Valley as far south as Eugene, Oregon. Figure 2.7 shows probably the largest erratic boulder in the region. This erratic

Figure 2.4 Dry Falls in Grand Coulee (panoramic photo by Rick Thompson). Dry Falls is 400 feet (120 meters) high and 4 miles (6 kilometers) long.

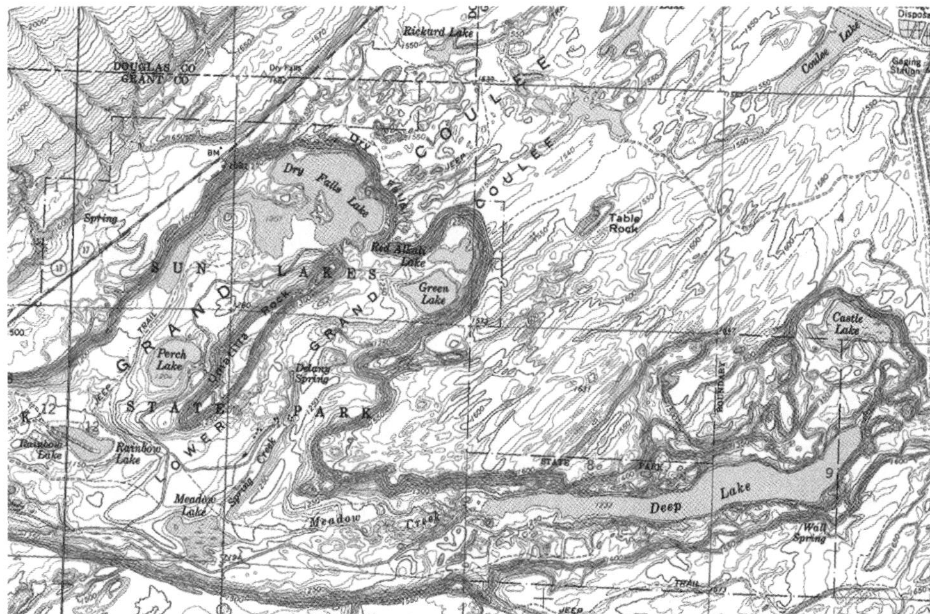

Figure 2.5 Topographic map of Dry Falls in the Grand Coulee (courtesy of the United States Geological Survey).

Figure 2.6 Large basalt boulders in Frenchman Coulee.

boulder, lying on a ridge about 30 miles south southeast of Arlington, Oregon, has been broken from one large boulder that was originally about 30 x 20 x 15 feet (9 x 6 x 4.5 meters). Granite erratics are embedded within rhythmites in the valleys of eastern Washington. A rhythmite is a repeating vertical sequence of two or more sediment types in a particular order (Bates and Jackson, 1984, p. 432). It is doubtful these erratics could have rolled into the rhythmites; iceberg rafting seems to be the only possibility. Further evidence for the iceberg origin of these erratics comes from debris lenses in the rhythmites. In one location near Sawyer, Washington, located in the Yakima Valley, I found a lens of cobbles and finer debris within the rhythmites (Figure 2.8). This material resembles an iceberg dump structure, produced when an iceberg becomes unstable, turns over, and dumps rocks and finer debris into the water that forms a mound on the bottom of the lake or ocean (Ovenshine, 1970).

Just in the Willamette Valley alone, Ira Allison back in 1935 reported 249 localities with erratic boulders (Figure 2.9). They are found strewn on the surface up to an elevation of 400 feet (120 meters) above sea level (ASL) and are composed of granite, granodiorite, several varieties of quartzite, schist, gneiss, and many other lithologies. Most of these rock types do not outcrop anywhere near this valley. Some very distinctive lithologies are also present, including a six-inch (15 centimeter) cobble of kyanite found in an excavation in downtown Portland that appears to be from near Revelstoke, British Columbia, and a group of baculite fossil-bearing rocks found near Gladstone, Oregon,

Figure 2.7 Erratic granite boulders, probably broken from one large boulder originally about 30 x 20 x 15 feet (9 x 6 x 4.5 meters), lying on a ridge about 30 miles (48 kilometers) south southeast of Arlington, Oregon.

Figure 2.8 Iceberg dump debris in rhythmites near Sawyer, Washington.

that outcrop in the intermontane valleys of British Columbia (Allen, Burns, and Sargent, 1986, p. 174). The largest meteorite ever found in the United States was discovered in 1902 two miles northwest of West Linn, Oregon. This meteorite in now believed to have been deposited by an iceberg during the Lake Missoula flood (Allen, Burns, and Sargent, 1986, p. 182). A replica of it is on display in front of the West Linn library (Figure 2.10).

An especially interesting large erratic is the Belleview Erratic in the Willamette Valley, just north of Oregon highway 18, halfway between McMinnville and Sheridan and southwest of Portland (Allen, Burns, and Sargent, 1986, pp. 182, 183). Before tourists chipped away pieces of this erratic, it weighed 160 tons (145,000 kilograms) with dimensions of 21 x 18 x 5 feet (6 x 5 x 1.5 meters)! Now there are only 90 tons (82,000 kilograms) remaining (Figure 2.11). The boulder is composed of argillite, a slightly metamorphosed shale. Argillite does not outcrop anywhere close to this area and is very similar to that found in the Belt Supergroup of northern Idaho and western Montana. This boulder could not have rolled hundreds of miles into place since the shale would not have survived intact. In fact, it probably could not have rolled more than several miles (5 kilometers). Furthermore, it is angular, indicating no smoothing by running water. The only possible mechanism of emplacement is from an iceberg raft. Bretz was completely baffled by all these erratic boulders.

A PLEXUS OF YOUTHFUL CHANNELS

After leaving the University of Washington for the University of Chicago, where he spent the remainder of his academic career, Bretz continued his field trips into the Pacific Northwest. Many of these field trips were on foot. He was still trying to understand the unique landforms in eastern Washington.

Eastern Washington is a broad, low elevation area, called the Columbia Basin or the Columbia Plateau that lies east of the Cascade Mountains (see Figure 1.1). Its surface rocks are composed of a series of basalt lava flows that were extruded from southeast Washington and northeast Oregon. The combined lava flows are up to 2 miles (3 kilometers) thick and cover an area of 64,000 mi^2 (164,000 km^2) (Tolan *et al.*, 1989). Unconsolidated silt deposits that are believed to be loess lie on top of the basalt in various places, mainly at higher elevations. The silt is called the Palouse Formation or Palouse loess.

Bretz noticed that from the Columbia River eastward, the silt and basalt has been deeply dissected into diverging and converging canyons or coulees with *vertical* walls

(Figure 2.12). The canyons form a plexus, generally having the same width and depth throughout any one channel. The walls of most coulees are vertical and 100 to 400 feet (30 to 120 meters) high, indicating their youthfulness (Bretz, 1923a, p. 589). Old canyons have a v-shape that increases in width downstream.

The bottoms of many coulees are flat, but many others consist of a rough butte-and-basin topography with little or no soil and little vegetation. There are also rock basins, potholes, and large abandoned waterfalls within the coulees (Baker, 1978b, p. 59). The rock basins are often long, narrow, and overdeepened. They are sometimes filled with lakes. The longest overdeepened rock basin is Rock Lake, 40 miles (60 kilometers) southwest of Spokane. It is 7 miles (11 kilometers) long and 100 feet (30 meters) deep (Baker, 1978c, p. 99). There are about 100 of these mostly dry scabland coulees in eastern Washington (Baker, 1978c, p. 82). Very few places on Earth possess such unique landforms (Bretz, 1959, p. 7). This area is called the Channeled Scabland. There are two large scabland complexes or groups of coulees (not including the Grand Coulee and Moses Coulee to be discussed below): the Cheney-Palouse scabland tract that forms the southeast boundary of the Channeled Scabland and the Telford-Upper Crab Creek tract (see Figure 1.1). The Cheney-Palouse tract is the largest and begins just southwest of Spokane, Washington, while the Telford-Upper Crab Creek complex starts just southwest of the confluence of the Spokane River with the Columbia River.

It is now surmised that such channels and overdeepened basins were carved by what is called a kolk in the context of a monstrous flood (Baker, 1978b, pp. 73-76; Alt 2001, pp. 41-43). A kolk is a very strong vertical vortex that develops within deep, super-fast flowing water. It is similar to a tornado but underwater. Dutch engineers have witnessed kolks lift and twirl riprap blocks as heavy as automobiles and carry them off downstream. So, it would be easy for these kolks to pluck the jointed basalt and carry large chunks of basalt downstream. Bretz (1924b) even anticipated the discovery of kolks by comparing the origin of the Channel Scabland to the overdeepened pools of the Columbia River channel at The Dalles, Oregon. The Columbia River has carved rock basins down to 115 feet (35 meters) below sea level at the head of Five Mile Rapid and 150 feet (45 meters) below sea level at the Big Eddy. He theorized that these deep pools were excavated during normal floods on the Columbia River. Hanging tributary valleys are seen at the tops of many of the dry coulees. They look like gutter spouts, where streams used to flow and sometimes still flow (Allen, Burns, and Sargent, 1986, p. 47). Hanging valleys indicate that the plexus of channels was carved rapidly, leaving previous drainage channels hanging. One of the most unsettling deductions that occurred to Bretz was the significance of hanging valleys in the *thick, soft* Ringold Formation near Hanford Washington, referred to in Chapter 1:

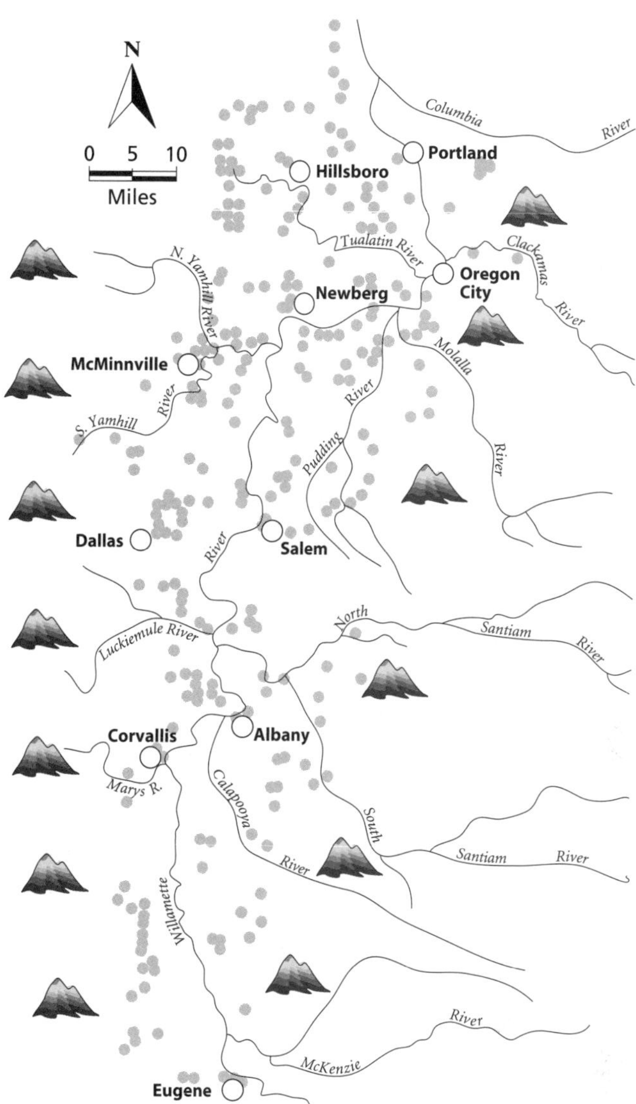

Figure 2.9 Plot of erratic boulders from the Willamette Valley (redrawn from Allison, 1935, by Mark Wolfe).

> These channels mouth in the wall of Columbia Valley from 150 to 400 feet [45 to 120 meters] above the river. The entire wall is of the weak Ringold sediment. Basalt here is below river-level. The Ringold never could have maintained the lip of a cataract at the mouths of the valleys. The writer's interpretation has been that the surface of the glacial Columbia [the Spokane flood] was as high along this wall as these channels now hang...A river at least 400 feet [120 meters] deep is therefore demanded for the highest channel mouth (Bretz, 1928a, p. 200).

No normal river or waterfall could maintain a height of 400 feet (120 meters) in such soft sediments; it would rapidly cut down through the soft sediment to the level of the Columbia River. Bretz's insightful but perplexing deduction of a flood at least 400 feet (120 meters) deep at this location was quite unsettling to him as well as to those who read about it in his publications.

THE GRAND COULEE

By far the largest and most difficult scabland channel to explain is the Grand Coulee. The coulee is 50 miles (80 kilometers) long and up to 1000 feet (300 meters) deep. There is a northern and a southern half separated by the broad Hartline Basin and Dry Falls. The origin of this coulee is still problematic today. Bretz intensely studied the Grand Coulee, even writing a monograph on the subject (Bretz, 1932). He concluded that a gigantic flood also carved it.

Many of the unique erosional features of the scabland channels are exemplified in the Lower and Upper Grand Coulee (Bretz, 1932; Baker, 1989a). The head of Upper Grand Coulee actually starts at a saddle in an east-west ridge, just south of the Columbia River. It is now known that the Lake Missoula flood excavated the large saddle after raising glacial Lake Columbia and overtopping the ridge. The base of the eroded saddle in the ridge is 550 feet (170 meters) *above* the Columbia River! Before excavation, the ridge was about 2400 feet (730 meters) ASL, or 1400 feet (425 meters) above the Columbia River (Baker and Bunker, 1985, p. 33). The flood lowered the ridge in this spot forming a saddle 860 feet (260 meters) below the original altitude of the ridge. The question Bretz tried to answer was: Why would a huge canyon start perpendicular to a ridge and not be excavated down to the level of the Columbia River? The only answer would be a short intense flood, which was an unacceptable conclusion for the geological establishment.

The reason the water was at least 1400 feet deep in this location is because another ice-dammed glacial lake filled the basin of the Columbia River, which was blocked by a lobe of the Cordilleran Ice Sheet (see Figure 1.1). The Spokane Flood quickly raised the level of this glacial lake, where the water overtopped the ridge at the head of the Upper Grand Coulee. Not surprisingly, this lake has been named glacial Lake Columbia. Extensive evidence for this glacial lobe is found on the Waterville Plateau west of the Upper Grand Coulee. The Waterville Plateau is the high area between the Grand Coulee and the Columbia River to the west. A large terminal moraine, giant erratic boulders, striated surfaces, eskers, and "haystack" erratics plucked from the basalt are north of a line from Lake Chelan to near Coulee City. Further support for this area being covered by an ice lobe is found on the Columbia River. From the notch in the east-west ridge at the north end of the Upper Grand Coulee to

Figure 2.10 Replica of a meteorite from near West Linn, Oregon.

Figure 2.11 Belleview erratic from Willamette Valley, Oregon. This large boulder is composed of a slightly metamorphosed shale, called argillite, which could not have rolled hundreds of miles from its nearest source in northern Idaho or western Montana.

just downstream of Wenatchee, the river is *not* part of the Channeled Scabland (Bretz, 1930a). If any place should be a scabland channel, it should be this stretch of the Columbia River–unless protected by ice that filled the whole channel and overflowed south onto the Waterville Plateau.

It is interesting that the floor of the upper Grand Coulee does *not* slope downstream like a normal river valley; it slopes northward and is lowest at the *northern* end! This indicates that a normal south-flowing stream or river did not erode the Upper Grand Coulee. Lake deposits with what are believed to be varves, silt/clay couplets deposited in one year, occupy the surface of the northern Grand Coulee. These "varves" indicate that the northern Grand Coulee was part of glacial Lake Columbia after the Lake Missoula flood had ended (Atwater, 1987). Otherwise, another major Lake Missoula flood would have scoured these lake sediments. Glacial Lake Columbia, therefore, survived the Lake Missoula flood or floods. The water that overtopped the east-west ridge at the head of Upper Grand Coulee likely flowed at first as a monstrous river to near Coulee City. It is believed that a huge waterfall developed at Coulee City and receded northward 25 miles (40 kilometers) forming the excavated Upper Grand Coulee. Where the waterfall began near Coulee City, Bretz found deep plunge pools up to 300 feet (90 meters) deep filled with gravel (Bretz, 1959, p. 20).

Figure 2.12 Coulee with vertical walls and generally flat bottom located just above the Potholes Cataract (photo courtesy of Michael Shaver).

Farther south, the Lower Grand Coulee was eroded along the southeast edge of the Coulee Monocline where the lava beds dip up to 45 to 60 degrees to the southeast from the flat-lying basalt of the Waterville Plateau to the west. A monocline is a local steepening of otherwise horizontal or uniformly dipping rock layers. One can see remnants of the eroded basalt of the monocline that stick out like giant fins in Lenore Lake, Lower Grand Coulee. These remnants are called hogbacks, narrow ridges of steeply dipping rock layers. When the monocline formed, the bending caused the basalt to weakened and crack, allowing the water of the flood to rapidly erode the lower portion of the monocline. Still, it would take a torrent of water to carve such a deep canyon. Bretz concluded that a receding waterfall ending at Dry Falls when the flood ended also formed the upper end of Lower Grand Coulee.

Hanging valleys are common along the sides of Grand Coulee. Baker (1989a, p. 52) describes the origin of these hanging valleys:

> Pre-flood drainage ravines, which descended the monoclinal limb, were left as hanging valleys high above the present valley

Figure 2.13 Upper Moses Coulee, Washington, at Jameson Lake.

Figure 2.14 Lower Moses Coulee, Washington, from the expansion bar at its entrance.

Figure 2.15 Hanging valley in Moses Coulee, Washington, near Jameson Lake.

floor, giving the skyline a gabled appearance

These hanging valleys represent valleys carved in the silt and basalt before the flood.

MOSES COULEE

Moses Coulee (see Figure 1.1) has been a bit of a mystery in its relationship to the Lake Missoula flood (Alt, 2001, pp. 99-108). It starts at an east-west terminal moraine in the middle of the Waterville Plateau and continues southwest to the Columbia River just south of Wenatchee. It is 40 miles (64 kilometers) long and 200 feet (60 meters) deep at its upper end (Figure 2.13) and up to 900 feet (275 meters) deep at its lower end, where it cuts through the Badger Mountain anticline (Figure 2.14). An anticline is a fold in the rock layers, generally convex upward, whose core contains stratigraphically older rocks (Bates and Jackson, 1984, p. 23). Because of its unique location, some scientists have thought Moses Coulee was carved by glacial streams during the melting of the ice lobe on the Waterville Plateau and is not related to the Lake Missoula flood.

However, Moses Coulee is a miniature Grand Coulee with vertical walls and a flat bottom that parallels the Grand Coulee only 10 miles (16 kilometers) to the west. Alt (2001, pp. 100-101) notes that the amount of talus and the degree of weathering of the rocks places Moses Coulee at the *same* age as the Grand Coulee and other scabland coulees. There are also numerous hanging valleys along the sides of Moses Coulee (Figure 2.15) similar to the Grand Coulee. Bretz considered Moses Coulee part of the Spokane flood (Bretz, 1923a, pp. 600-602). Others who have examined it since also agree with his conclusion (Hanson, 1970). Glacial till from the Withrow moraine on the Waterville Plateau lies above the Lake Missoula flood deposits, indicating that the edge of the Cordilleran glacial lobe surged southward after the flood. For Moses Coulee to form, the Grand Coulee must have been nonexistent, because the water would have been diverted down the Grand Coulee before reaching the head of Moses Coulee.

Based on this information, it appears that Moses Coulee was carved first when the waters of the Spokane flood first overflowed into glacial Lake Columbia from the east, raising the level of the lake. At the beginning of the Spokane flood, water poured out of Lake Columbia along the edge of the ice sheet. From there it traveled generally west along the edge of the ice on the Waterville Plateau, and then southwest away from the ice starting in a low area of the Waterville Plateau, rapidly cutting Moses Coulee (Hanson, 1970). After Moses Coulee was formed, Grand Coulee would have been initiated resulting in a new channel forming and diverting the water away from Moses Coulee.

WATER-SHAPED SILT HILLS

Within the wider coulees and in the large scabland tracts, Bretz noticed even more evidence of a water origin for the Channeled Scabland. The Palouse silt was not completely eroded in many scabland areas, especially within the Cheney-Palouse scabland tract. In fact, the silt remnants form lens-shaped hills. The long axes of the silt hills are three times longer than their widths and lie parallel to the flow direction of the flood. The silt hills have the appearance of the prow of a ship converging in the upstream direction of flow. The long sides of the hills are quite steep, often with a slope around 30^0 to 35^0 and are usually ungullied (Bretz, 1928a). Only narrow channels separate many hills or hill groups, while some hills stand alone in the middle of scabland. There are hundreds of these lens-shaped silt "islands," some of them up to 200 feet (60 meters) high (Baker, 1978c, pp. 88-91) adorning the flood path (Figure 2.16). Bretz thought most of these were islands in the flow, but based on the maximum flood height, most of them had to be eroded and shaped *underwater*.

The soft streamlined silt islands stand in stark contrast with the scoured, potholed rock between them. We are left with the question as to why the totally flooded silt hills were not completely eroded, while the channels around the hills cut down into the much harder basalt. It appears to be a mechanical paradox, but Baker explains it by using the physics of water flow. He believes the silt hills are "...equilibrium forms, elongated sufficiently to reduce pressure drag, but not so long that they create excessive skin resistance" (Baker, 1978c, p. 90).

GIANT GRAVEL BARS

Bretz was amazed at the size of the gravel deposits in the Channeled Scabland and wrote often about them (Bretz, 1928b; 1929a, b). Based on their location and the other features of these huge gravel mounds, he deduced that they were *river bars*, but not from ordinary rivers. This meant that the flood had to be at least as high as the top of the bar (Bretz, 1928d)–a straightforward deduction that scientists of the time could not comprehend.

Many gravel bars are found along the flood path. There are three general types of bars in the Channeled Scabland: 1) pendant or longitudinal bars that formed downstream from an obstacle, 2) expansion bars that developed as the flow spread out and slowed after passing through a narrow constriction, and 3) eddy bars formed at the mouths of tributary valleys due to two converging flood flows or a swirling eddy.

The pendant bar is the most common gravel bar in the Channeled Scabland (Figure 2.17) (Baker, 1973). The locus of bar initiation may be a rock projection, the downstream end of a silt hill, or even the bend of a pre-flood meander-

Figure 2.16 Lens-shaped silt hill in the Cheney-Palouse scabland tract just west of Palouse Falls.

Figure 2.17 Pendent gravel bar at Malden, Washington.

Figure 2.18 Foresets or dipping gravel layers in a pendent bar just west of a ridge about 30 miles (48 kilometers) south southeast of Arlington, Oregon.

ing valley. The gravel in pendent bars was often deposited in huge foresets which are beds of gravel that dip downflow, like a gravel delta, and which is typical of flowing water (Figure 2.18). The gravel in some of these bars occasionally consists of large basalt blocks, many still having a polygonal columnar structure. In large river valleys such as the Snake River, the foresets dip *upvalley*, east of the confluence with the Palouse River, indicating a torrential flow of great depth and vigor rushing *up* the river (Bretz, 1929b, p. 509). Similar to the streamlined silt hills, the bars are elongated no more than about three times their width, which appears to be an equilibrium form (Baker, 1978c, p. 94).

Eddy bars block nearly every pre-existent tributary valley entering the scabland tract from the east (Baker, 1978c, p. 95). Foresets in the bars at the mouth of these tributaries often dip up the tributary indicating upvalley flow (Bretz 1929a, b). These tributaries contain exotic rocks and stratified silt, called rhythmites–further proof of upvalley currents. Eddy bars are sometimes observed high up along the sides of short valleys along the flood path. A good example of this type of eddy bar is found across the Columbia River northwest of The Dalles, Oregon (Figure 2.19 and 2.20). This is powerful evidence for a deep catastrophic flood.

The huge expansion bars that Bretz discovered are also impressive features of the flood tract. In the eastern Cheney-Palouse scabland track, these bars are rather small, covering areas only up to 20 mi^2 (50 km^2) (Patton and Baker, 1978, p. 126). But, the Ephrata Fan that covers much of the Quincy Basin is the most impressive expansion bar in the western scablands. It is about 600 mi^2 (1,500 km^2) in area and up to 130 feet (40 meters) thick (Bretz, Smith, and Neff, 1956, p. 969-974). It formed as water exited the lower Grand Coulee at a peak velocity of about 65 mph (30 m/sec) (Baker, 1978b, p. 65). This fast flow dumped all the eroded rocks from Grand Coulee onto the broad Quincy Basin and formed the Ephrata Fan. The fast flood flow also scoured out a rock basin within the constriction, now occupied by Soap Lake, and piled the rock debris downstream to the south 250 feet (75 meters) higher than the surface of the lake (Bretz, 1932, p. 19)! Soap Lake is about 300 feet (90 meters) deep (Bretz, Smith, and Neff, 1956, p. 969). So the vertical distance between bedrock at the bottom of the lake and the top of the fan is about 500 feet (150 meters)! The surface of the Ephrata Fan is lobate with channels carved on top that are sometimes filled with linear lakes. Moses Lake is a good example. The top of the fan is strewn with boul-

Figure 2.19 Eddy bar in a small valley north across the Columbia River from The Dalles, Oregon.

ders, some quite large, of which 95% are composed of basalt and about 5% are granodiorite (Rice and Edgett, 1997, p. 4190). The nearest source for the granodiorite is about 40 miles (65 kilometers) upstream at the northern end of the upper Grand Coulee. Another impressive expansion bar is the gravel bar of Moses Coulee (see figure 2.14). It is more than 300 feet (90 meters) thick at its confluence with the Columbia River, but covers a much smaller area than the Ephrata Fan (Bretz, 1923b, p. 646; Bretz, 1930a).

Figure 2.20 Close-up view of the eddy bar at The Dalles, Oregon. The linear shoreline-like features could be flow levels of the water that either were faster or longer lasting.

THE HUGE PORTLAND "DELTA"

The most impressive expansion bar is the Portland "Delta" in the Portland-Vancouver area (Figure 2.21) (Bretz, 1925b; 1928d, pp. 696-700). The gravel bar is 200 mi^2 (500 km^2) and more than 350 feet (110 meters) deep! From hundreds of exposures, Bretz noted that all of the gravel contains foresets dipping westward, indicating predominant westward flowing water. Fluvial flow features were noted on top of the bar and around several volcanic buttes. Rocky Butte in northeast Portland is a good example of a volcanic butte with fluvial flow features surrounding it. Bretz surmised that the Spokane flood rushed from the Channeled Scabland through the narrow Columbia Gorge at a high speed and then slowed down, depositing its load of sediment and rock.

We now know that after exiting the narrow constriction of the Columbia Gorge at Crown Point, the velocity was about 80 mph (35 m/sec) (Benito, 1997). The Columbia and Willamette Rivers flow in dissected troughs of the bar. Bretz postulated that the Portland Delta must have formed underwater when the Spokane flood flowed above the *highest* level of the bar. That would make the flood in the Portland-Vancouver area about

Figure 2.21 The Portland "Delta" or expansion bar (redrawn from Bretz, 1928d, p. 696 by Mark Wolfe).

Figure 2.22 Giant ripple marks in eastern Washington (photo by P. Weis courtesy of the U. S. Geological Survey).

Figure 2.23 Wallula Gap, Washington. This gap is one mile (1.6 kilometers) wide. The water could not exit through this gap fast enough and caused a temporary lake about 750 feet deep in the Pasco Basin.

400 feet (120 meters) deep! It is no wonder that his critics thought he had gone too far.

GIANT RIPPLE MARKS

Although not discovered until later by aerial photography, there are about 100 sets of giant ripple marks, predominantly on top of the gravel bars, along the flood path (Baker and Bunker, 1985, p. 18) (Figure 2.22). The ripples display remarkable symmetry of form (Baker, 1978c, p. 107): "Mean ripple heights for a ripple field are closely related to mean ripple chords..." The profile of the ripples is asymmetric in profile; the downstream face having a slope of 18 to 20° while the upstream face is smoother at 6 to 8°. The sediments in the ripples are gravel with a median size in the pebble fraction but containing rocks up to 5 feet (1.5 meters) in diameter (Baker, 1978b, p. 109).

A FLOOD SEVERAL HUNDRED FEET DEEP

Bretz was able to deduce that during the flood the depth of flow was several hundred feet (about 100 meters) or more. He was able to estimate the height of the flood at maximum by estimating *high water marks*. These gave him a fairly accurate measure of the flood's depth. Even more disconcerting to his critics, he also discovered that the water was much deeper at a few narrow constrictions, such as Wallula Gap and the Columbia Gorge where the water depth was about 750 feet (230 meters) deep. Wallula Gap is a one-mile (1.6 kilometer) wide gorge or water gap through the Horse Heaven Hills anticline of the Columbia River Basalt Group just south of the low basin around Pasco, Washington (Figure 2.23).

The height of erratics and scoured basalt and silt also reinforced his estimates of maximum water depth. Bretz further noticed that the flood crossed the ridge between the Snake River and Washtucna Coulee. The ridge parallels the Snake River Valley and is about 10 miles (16 kilometers) wide (Bretz, 1923b). It breached the ridge at two locations and cut canyons 500 feet (150 meters) deep (see Chapter 9).

In 1925, Bretz (1925a, b) published evidence that the flood could not have drained through Wallula Gap and the Columbia Gorge fast enough to prevent ponding to the north. As a result the ponded water formed a large temporary lake that backed up through south central Washington, and spread *up* the Snake River past Lewiston, Idaho, well over 100 miles (160 kilometers) upriver. Based on high water marks, he deduced that the water was about 750 feet (230 meters) above the present day Columbia River at Pasco, Washington, and

Arlington, Oregon, downstream from Wallula Gap, tapering to about 430 feet (130 meters) above the river at Portland, Oregon.

Once he determined the high water marks, he was able to use standard hydraulic equations to estimate the maximum flood flow. Using Chezy's equation, relating the flow velocity to the maximum height of the water and channel parameters, he was able to estimate the amount of water and the velocity of the flow through Wallula Gap (Bretz, 1925b, p. 258). Because he assumed too gentle of a water-surface slope and uniform flow, he significantly underestimated the flow (Baker, 1973, p. 21). Nevertheless, Bretz's estimates were catastrophic and deep enough to raise eyebrows.

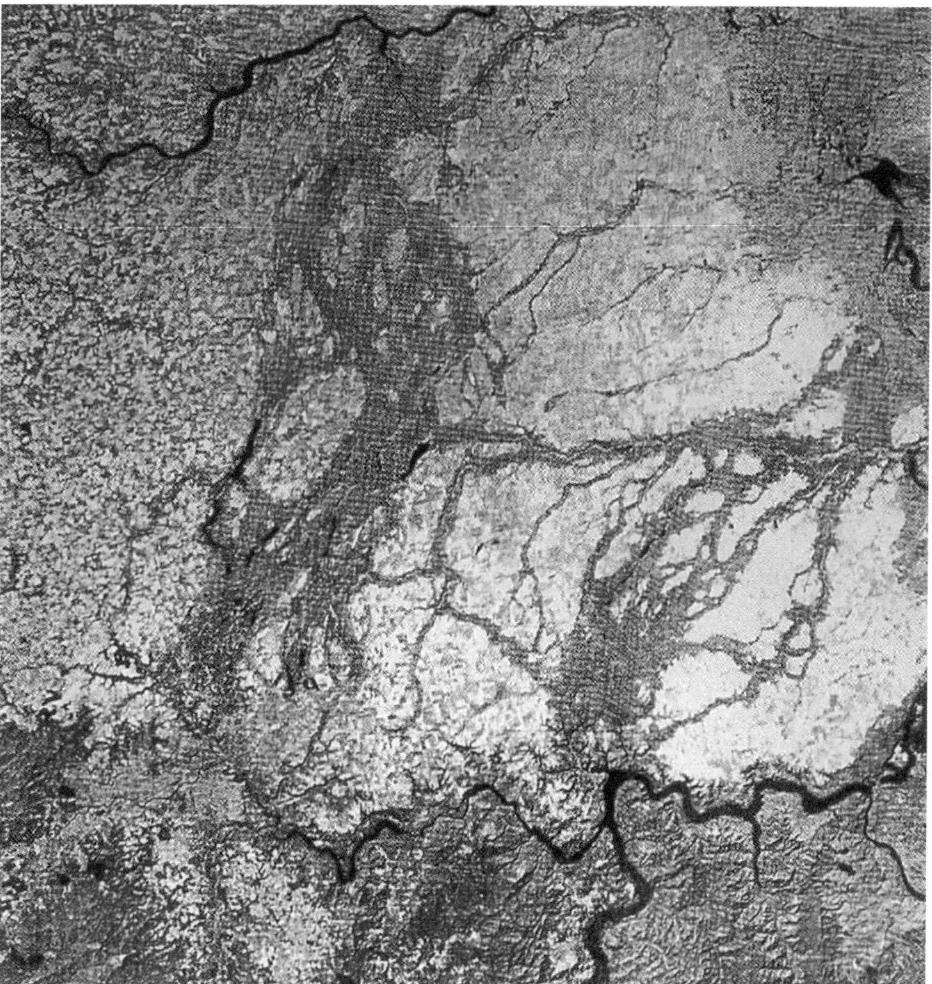

Figure 2.24 Landsat picture of the Channeled Scabland.

WATER IN ALL THE CHANNELS AT ONCE!

It was difficult for Bretz's critics to accept a flood several hundred feet (about 100 meters) deep raging down one channel, but possibly they could explain it by meltwater from a shifting ice sheet or a burst ice jam or some such mechanism. However, Bretz realized that the heads of all the scabland tracts opened up to the north or northeast at about the *same* altitude of 2,350 to 2,550 feet (715 to 780 meters) ASL (Bretz, 1923b, 1969, p. 527). Furthermore, he noted that all four exits from Quincy Basin developed at the *same* altitude, implying a completely flooded basin (Bretz, 1923a). (Besides the three erosional cuts in the basalt ridge along the western Quincy Basin, the fourth outlet was to the southeast through Drumheller Channels.) Bretz came to the shocking conclusion that all the tracts and other minor scabland channels must have been flooded *at the same time*!

Bretz's flood, composed of diverging and converging channels, was about 100 miles (160 kilometers) wide! The outline of all these flooded channels can be seen very clearly from space (Figure 2.24). The dark scabland channels show up against the light colored, cultivated, non-eroded to partially eroded silt. Based on divide crossings and other high altitude erosional and depositional features, Bretz determined that at least one Spokane flood was hundreds of feet (about 100 meters) deep and as wide as a hundred miles (160 kilometers)–an enormous magnitude! It is no surprise that the overwhelming reaction among the geological community was one of shock and disbelief.

ORIGIN OF SPOKANE FLOOD UNKNOWN AT FIRST

For about 10 years, Bretz did not know the origin of the floodwater that carved the Channeled Scabland. In essence, he had suggested a sensational hypothesis, based on copious field data, *without a source*. This did not bode well for the acceptance of his hypothesis, and several geologists jumped on this weakness as proof there never was such a flood.

Bretz at first hypothesized that a marine submergence caused the unique features of eastern Washington, the Columbia Gorge, and the Willamette Valley (Bretz, 1919)–an hypothesis that had been suggested by geologists before him. He soon changed his mind. It was not long before he realized the Channeled Scabland was formed by flowing water that originated from somewhere north or east of Spokane. He then weakly postulated that the flood was either due to a very rapid, short-lived climate amelioration during the ice age or from a gigantic meltwater pulse caused by a volcanic eruption beneath the Cordilleran Ice Sheet to the north (Baker, 1978a, p. 9).

Bretz should have known the source of the water for the flood from the beginning. It should have been no mystery at the time, since the existence of a monstrous ice-dammed lake was known to lie just to the east of the Channeled Scabland at the peak of the ice age.

Chapter 3

Glacial Lake Missoula and Its Catastrophic Breach

Driving through the beautiful mountain valleys of western Montana, it is hard to imagine that a vast lake had once filled these valleys (Figure 3.1). The existence of this lake was first recognized over 100 years ago, even *before* Bretz postulated his enormous flood through eastern Washington. In the late 1800s renowned geologist, T.C. Chamberlin noted a series of parallel faint "watermarks" on the hills in the Flathead Lake region (Pardee, 1910, p. 376; Chambers and Curry, 1989, p. 4). Chamberlin went so far as to suggest that the watermarks were caused by a lake that was formed by an ice dam in the Lake Pend Oreille region of northern Idaho. He was remarkably accurate, but in pointing out this lake he was ahead of his time by three-quarters of a century. But Chamberlin failed to consider the consequences of the inevitable breaching of such an ice dam. Only Bretz recognized the potential of glacial Lake Missoula, as it was called, to form the Channeled Scabland, but it eluded even him for about ten years.

LAKE MISSOULA – A TEMPORARY GREAT LAKE OF NORTH AMERICA

The Great Lakes of North America are among the largest lakes in the world and collectively are the world's largest body of fresh water. They occupy the border area between the north central United States and adjacent Canada. During the ice age another Great Lake, glacial Lake Missoula, developed when water ponded behind an enormous ice dam in the Lake Pend Oreille area of northern Idaho, just as Chamberlin recognized. Lake Pend Oreille is an overdeepened lake that lies in the southern part of the Purcell Valley, a valley that stretches north into British Columbia. Lake Pend Oreille is 1000 feet (300 meters) deep (Breckenridge, 1989, p. 19). Overdeepened valleys, lakes, and fjords are common in the formerly glaciated areas of the high and mid latitudes of the Northern

Figure 3.1 Glacial Lake Missoula, western Montana, at maximum volume (drawn by Mark Wolfe).

Hemisphere.

The Purcell Valley lies between the Selkirk Mountains on the west and the Cabinet Mountains on the east. These mountains rise about 4000 to 5000 feet (1200 to 1525 meters) above the valley floor. Some of the peaks in these ranges were not glaciated during the ice age (Walker, 1967). This is typical at the edge of ice sheets where the lower areas–valleys, plains–were glaciated. The ice in the Purcell Valley must have been between 2500 and 5000 feet (760 and 1525 meters) deep during the height of the ice age.

What Chamberlin called watermarks in the Flathead Lake region are remnant shorelines of glacial Lake Missoula and are observed elsewhere on the mountain sides of western Montana. The shorelines are easily observed east and north of Missoula on Mount Sentinel (Figure 3.2) and Mount Jumbo (Figure 3.3). Shorelines are also etched on the hills above the Little Bitterroot Valley, 75 miles (120 kilometers) northwest of Missoula (Figure 3.4) and on the Polson Moraine, just south of Flathead Lake (Chambers, 1971, pp. 13, 15). They are found south of Missoula along the edges of the long north-south Bitterroot Valley as well, showing up mainly on the north and northwest facing slopes, especially where the valley widens (Weber, 1972, p. 24). The locations of the shorelines in the Bitterroot Valley suggest north to west winds, similar to the prevailing winds today.

Figure 3.2 Shorelines on Mount Sentinel, just east of Missoula, Montana (photo by P. Weis courtesy of the U.S. Geological Survey).

Figure 3.3 Shorelines on Mount Jumbo, just northeast of Missoula, Montana.

Shorelines and beaches have been notched on nested terminal moraines in the Bitterroot Mountains west of the Bitterroot Valley (Weber, 1972). The moraines were formed by mountain ice caps along the highest peaks of the Bitterroot Mountains that drained eastward down the valleys, reaching the floor of the Bitterroot Valley south of Hamilton. The juxtaposition of the terminal moraines with the shorelines is another indication that the lake was associated with the ice age. Besides shorelines, high deltas associated with tributary valley streams point to the existence of an ancient lake. High deltas exist in the Bitterroot Valley (Weber, 1972), but they are generally rare and small elsewhere in western Montana (Pardee, 1910, 1942). Such isolated, small deltas indicate that glacial Lake Missoula was short lived.

Exotic or erratic boulders high up on the sides of the surrounding mountains attest to iceberg rafting in this lake (Pardee, 1910). Weber (1972, p. 24) noted that erratic boulders are abundant on the shorelines in the Bitterroot Valley. They are clearly iceberg deposited since their parent rock is not found anywhere in the vicinity. Interestingly, they are always found *below* the elevation of the highest shoreline, further pointing to their origin by iceberg rafting on glacial Lake Missoula.

Lake deposits locally lie on the valley floor of the former lake (Figure 3.5). They are often interrupted by rhythmically bedded silt and clay rhythmites. Some investigators believe these rhythmites are varves, a silt/clay couplet deposited in a lake during a one-year period. In varves, the coarse silt sublayer is thought to have accumulated over the summer while the clay sublayer is believed to have slowly settled out in winter after ice formed on the lake. Most of the main valley lake deposits were scoured out when Lake Missoula emptied, just as one would expect from a catastrophic drainage of the water. This could be why there are no significant lake deposits from glacial Lake Missoula in the larger tributaries of the Blackfoot, Bitterroot, and Clark Fork Rivers east and south of Missoula (Pardee, 1942, p. 1579; Weber, 1972, p. 24; Chambers, 1984, p. 190). Fortunately, there are enough lake deposits preserved in protected areas north and west of Missoula to demonstrate the existence of the lake. The best location to view lake silts and "varves" is along Interstate 90 at the intersection of the Missoula and Ninemile Valleys about 30 miles (48 kilometers) west of Missoula (see Figure 5.9). This section plays a significant role in the controversy over how many times Lake Missoula filled up, as we will see in Chapter 4.

Figure 3.4 Shorelines along the eastern hills of the Little Bitterroot Valley, 75 miles (120 kilometers) northwest of Missoula, Montana.

Figure 3.5 Lake deposits in the Little Bitterroot Valley, northwest of Missoula, Montana.

The highest observed shoreline of glacial Lake Missoula on Mount Sentinel, just east of Missoula, is 4150 feet (1265 meters) ASL, or about 1000 feet (300 meters) above Missoula (Chambers and Curry, 1989, p. 4). In the Bitterroot Valley the highest shoreline is 4240 feet (1295 meters) ASL (Weber, 1972). The differences in altitude could be due to either preservational differences or variable tectonics. Aerial photography shows a distinct difference between the texture of the land below and above the highest shoreline (Weber, 1972, p. 27). Glacial Lake Missoula was 2000 feet (600 meters) deep in the lower Clark Fork River Valley near the Montana-Idaho border. Based on the highest lake shorelines, glacial Lake Missoula covered an area of about 3000 mi^2 (7700 km^2) and had a volume of about 540 mi^3 (2210 km^3). The volume of water was three times the volume of Lake Erie and one half the volume of the larger Lake Michigan. For a relatively short time glacial Lake Missoula was another one of the Great Lake in North America

THE DAM BURSTS

Joseph Pardee studied the geology of western Montana and early realized the existence of this enormous lake. At first, Pardee wrote that Lake Missoula emptied slowly through its ice barrier. He later changed his mind and published evidence for at least one catastrophic drainage. The peak of the ice age had already been reached so the ice dam in the Purcell Valley had begun to melt. Gradual melting of the surrounding ice would cause the lake to grow larger and larger while the ice thinned. It stands to reason that the dam would have

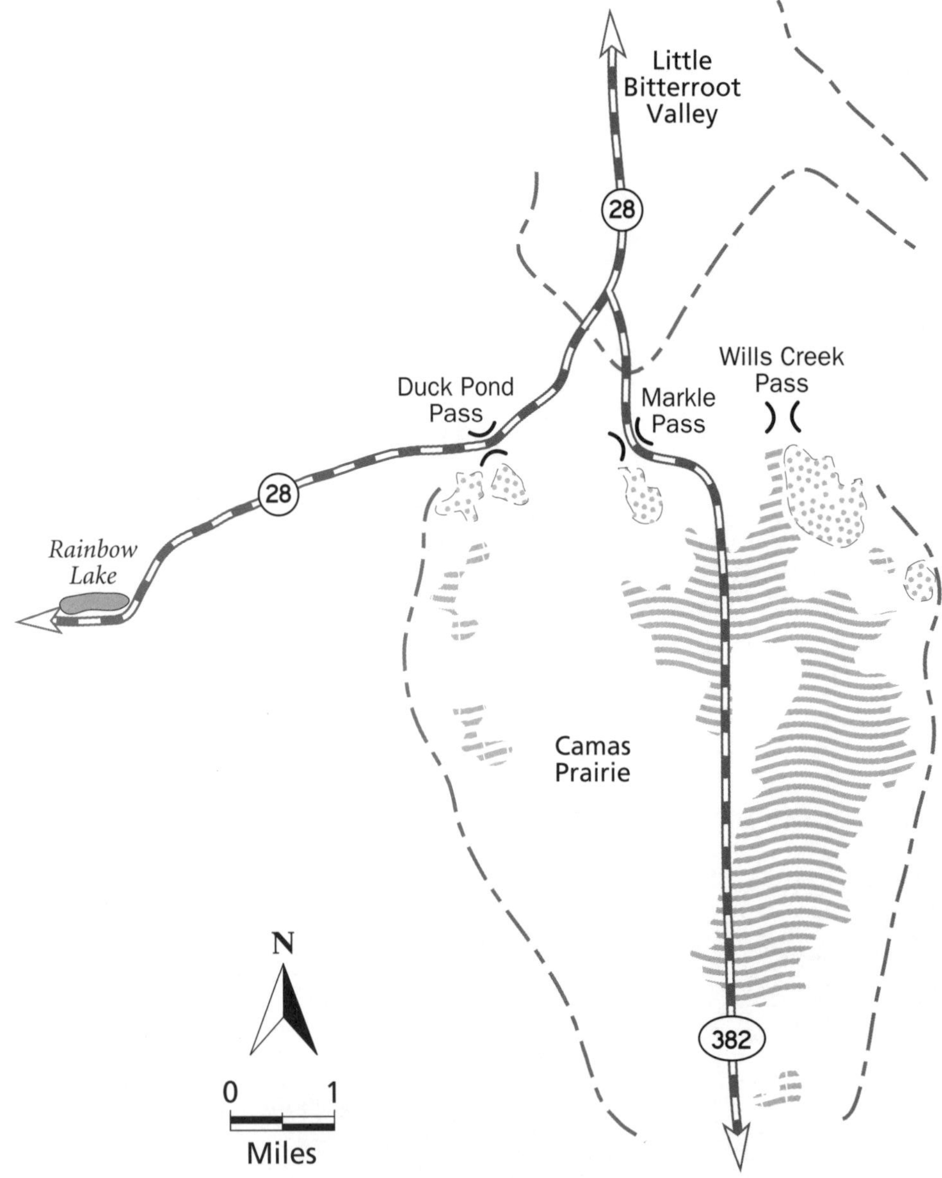

Figure 3.6 Map of Camas Prairie and vicinity showing water flow directions (redrawn and modified from Alt, 2001, p. 40 by Mark Wolfe).

burst suddenly, because of the pressure of the lake behind it, and because the water would eventually lift the ice barrier, since ice is less dense than water. Once the breach started, the increasing water flow would have catastrophically widened the breach. However, this reasonable scenario was too much of a catastrophe for the scientists of the day to accept.

Even within the area formerly occupied by glacial Lake Missoula, the evidence is impressive for a catastrophic emptying of the lake. The best evidence is in the Camas Prairie, a small intermontane basin 60 miles (95 miles) northwest of Missoula (Alt, 2001, pp. 37-40). This basin is 12 miles (19 kilometers) long and 6 miles (10 kilometers) wide (Figure 3.6). At its highest level, glacial Lake Missoula was 800 feet (245 meters) above the passes to the north and northeast of Camas Prairie. When the ice dam burst, the water rushed south at high speed over the low mountains and through the passes into Camas Prairie. The strong channeled flow scoured and plucked the rock through the passes leaving a scene much like the Channeled Scabland of eastern Washington (Figure 3.7). Markle and Wills Creek Passes, in particular, exhibit channeled excavations, discontinuous gravel deposits associated with projecting knobs, and gravel bars at the southern downstream end. This is indicative of linear scouring, mostly by plucking from kolks, fierce underwater whirlpools, similar to the action of tornadoes in the atmosphere (Alt, 2001, pp. 42-43).

As the lake water continued to rush down the north slope of Camas Prairie it formed giant gravel ripple marks. The gravel is composed of angular debris in the cobble-size range eroded from the surrounding mountains (Alt, 1987, p. 35). The ripple marks even contain some "varve" rip up clasts (Lister, 1981). The ripple marks range from a few feet to about 50 feet (15 meters) high and are spaced 200 to 500 feet (60 to 150 meters) apart (Pardee, 1940). They look like linear mounds on the ground (Figure 3.8) and are too large to be easily recognized as water ripples, but from the air they show up clearly as huge ripple marks (Figure 3.9).

Just as remarkable as the scoured passes and giant ripple marks of Camas Prairie is the huge gravel bar that extends from Rainbow Lake (Figure 3.10) over a divide west of Camas Prairie (Pardee, 1942; Alt, 2001, pp. 41-47). This bar is about 6 miles (10 kilometers) long and at its western and southern end forms a terrace 500 feet (150 meters) high (Figure 3.11). The bar is composed of angular rocks up to 6 feet (2 meters) long (Alt, 1987, p. 35). Rainbow Lake starts at the eastern end of the bar and is shaped like a two kilometer long plunge pool that would have been carved by the convergence of two currents from the passes to the east (Pardee, 1942, pp. 1385-1386).

Eddy gravel bars high up on side gulches in the lower Clark Fork Valley provide further evidence for the catastrophic draining of glacial Lake Missoula (Figure 3.12). One such eddy bar is located 300 feet (90 meters) above the river and is about 150 feet (45 meters) thick (Breckenridge, 1989, pp. 15,16). Pardee (1942, pp. 1591-1592) lists 32 of these high eddy bars. Foresets dip upstream in some of the eddy bars, indicating that they must have been formed in a giant back flow or eddy. The highest eddy bar in the Clark Fork Valley is about 1200 feet (365 meters) above the river. These high bars indicate a powerful current at least 1200 feet (365 meters) deep rushing down the lower Clark Fork River Valley, indicating that the ice dam totally failed!

THE CATASTROPHIC FLOOD

When glacial Lake Missoula's ice dam burst, a huge wall of water raced through eastern Washington and formed the Channeled Scabland. A person could have heard and felt the rumble of the coming flood for a half hour before it hit. After the existence of the Lake Missoula flood was finally accepted, it was up to Victor Baker (1973) to quantitatively refine Bretz's estimates of water depths and velocities in eastern Washington. Baker (1978b, p. 59) lists four methods for determining the high-water level at maximum flood discharge: 1) the highest eroded channel margin, 2) high-level depositional features, 3) the maximum elevation of ice-rafted erratics, and 4) divide crossings. By these methods, Baker (1973) estimated a peak water discharge of 740 million ft^3/sec (21 million m^3/sec) through the Spokane Valley, just downstream from the dam burst.

Figure 3.7 Scabland at Markle Pass, western Montana, between the Little Bitterroot Valley to the north and Camas Prairie to the south.

Baker's estimate was rather crude (Kiver, Stradling, and Baker, 1989, p. 26). So, the issue became rather contentious, largely because a number of scientists were uncomfortable with such a huge discharge. Much theoretical work was subsequently instigated to determine the mechanism and speed of dam bursting and the resulting flood discharge (Clarke, Mathews, and Pack. 1984; Beget, 1986a; Craig, 1987; Breckenridge, 1993). Some scientists applied a strict uniformitarian approach, based on modern examples from glacial lake outbursts, called jökulhlaups. Richard Waitt (1985, p. 1281) emphasized the uniformitarian approach of using these modern jökulhlaups as an estimate for the peak Lake Missoula flood:

> Most or all such [ice dammed] lakes drain before water rises enough to overtop the ice dam. There is neither field evidence nor theoretical reason that the huge glacial Lake Missoula, whose outlet was via the ice dam, should have behaved radically differently from small ice-dammed lakes.

The uniformitarian approach resulted in much less discharge than Baker (1973) proposed. The lowest estimated discharge was about 10% of Baker's estimate. Other models

Figure 3.8 The Camas Prairie, western Montana, ripple marks as seen from the ground.

produced multiple Lake Missoula floods that pleased many geologists.

However, the principle of uniformitarianism failed because field evidence for deeper water downstream was either ignored or minimized. When Craig's much smaller flood could not produce the high water marks around Spokane, he simply brushed off this evidence by claiming that the high water marks were caused by "...breaking waves far up the side of ridges along the Rathdrum Prairie" (Craig, 1987, p. 324). This was not accepted. Furthermore, glacial Lake Missoula held *one thousand times* as much water as the largest modern ice-dammed lake. Hence, modern examples of jökulhlaups *likely do not apply* with so much more water. This is called the scale problem (Baker, 1978c, p. 81). In other words, a larger volume of water will act differently than a small volume of water; one must be very careful in extrapolating data from small events to large events.

In the meantime Baker reworked his figures and came to the conclusion that his method of estimating discharge was about 30% too high (Baker and Bunker, 1985, p. 12). Later, O'Connor and Baker (1992) reanalyzed the data from around Spokane. The height of the maximum flood was estimated at about 2625 feet ASL with a discharge of at least 600 million ft^3/sec (17 million m^3/sec). These figures are generally accepted today. Thus the initial outburst from glacial Lake Missoula was 15 times the combined flow of all the rivers in the world (Baker and Bunker, 1985, p. 2). This discharge demonstrates that the ice dam in northern Idaho *failed completely* and that uniformitarian models using modern ice-dammed lakes are inappropriate. All 540 mi^3 (2210 km^3) of water from glacial Lake Missoula rushed into eastern Washington in only 3 days (O'Connor and Baker, 1992).

Figure 3.9 Oblique aerial view of giant ripple marks of the Camas Prairie, western Montana caused by the draining of glacial Lake Missoula (photo by P. Weis, number 24, courtesy of U.S. Geological Survey).

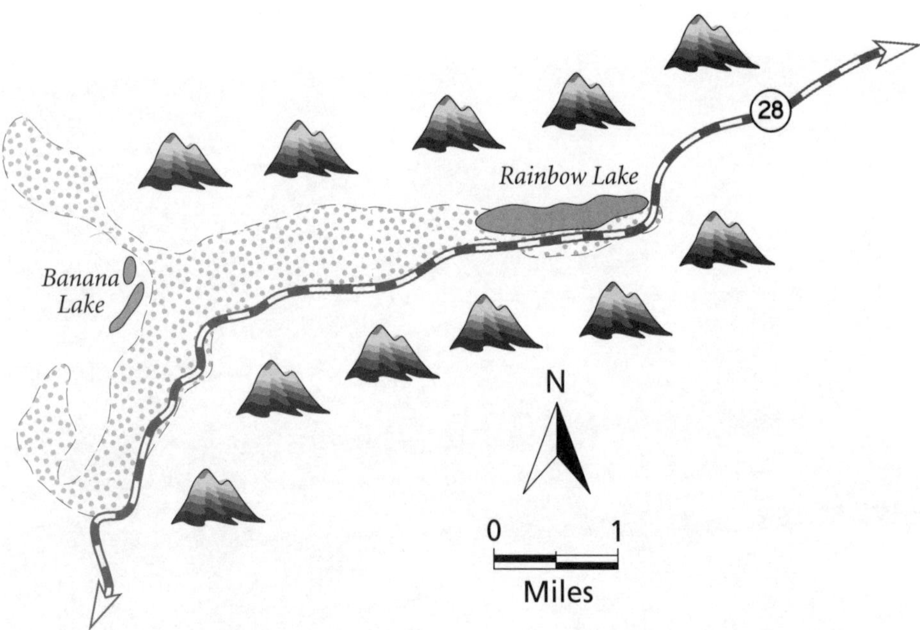

Figure 3.10 Rainbow Lake and gravel bar, over a small divide west of Camas Prairie (drawn by Mark Wolfe).

O'Connor and Baker (1992) also refined Bretz's original estimate of discharge through the narrow constriction at Wallula Gap. They calculated a maximum discharge through Wallula Gap of 350 million ft^3/sec (10 million m^3/sec) (O'Connor and Baker, 1992), significantly higher than Bretz's original estimate that shook the geological world at the time. Because all the water released by the ice dam could not flow fast enough through Wallula Gap, an immense lake around 750 feet (230 meters) deep developed. This temporary lake at maximum depth contained about 300 mi^3 (1230 km^3) of water, over half the original volume from glacial Lake Missoula. As the water rose in the temporary lake, it backwashed up the Yakima, Snake, and Walla Walla River Valleys. It is estimated that this lake lasted for about 5 days in southeastern Washington.

Baker (1973) also calculated peak velocities at a number of points along the flood route based on the high-water mark and channel geometry. He calculated mean flow velocities up to 65 mph (30 m/sec) at Soap Lake, southern Grand Coulee, and Staircase Rapids, just north of the town of Washtucna. Later calculations between Wallula Gap and Portland have yielded flow velocities up to 80 mph (35 m/sec) at Crown Point and Mitchell Point in the Columbia River Gorge (O'Connor and Waitt, 1995a, pp. 54-55). The flood was about 400 feet (120 meters) deep rushing through where Portland, Oregon, now exists, and as a result flowed south up the Willamette Valley as far as Eugene, forming another large, temporary lake. The maximum flood inundated 16,000 mi^2 (41,000 km^2) of land and eroded 50 mi^3 (200 km^3) of silt, sediment, and basalt (Allen, Burns, and Sargent, 1986, p. 3, 4).

Figure 3.11 The end of the giant gravel bar extending out from Rainbow Lake, western Montana.

Figure 3.12 Eddy bar in a side gulch along the Clark Fork River near Perma, Montana.

Chapter 4

How Many Lake Missoula Floods?

Once the Spokane flood became established, it was natural for uniformitarian scientists to ponder the question of whether there could have been more than one Lake Missoula flood. The number of Lake Missoula floods has occupied modern research for several decades, and has become a controversial subject. After the ice dam burst and Lake Missoula emptied, a wall of ice about 3000 feet (915 meters) high would have remained just north of the breach. This ice would surge southward and partially refill the breach, since ice flows like a plastic with a velocity proportional to surface slope. The vertical surface slope of the ice wall would cause an almost instantaneous southward acceleration. It is therefore likely that the ice dam reformed one or more times resulting in further lakes and floods at the peak of the ice age. It is also reasonable that such a scene could have repeated itself dozens of times, depending especially on the length of time the ice sheets remained at glacial maximum.

Figure 4.1 Bar in Snake River Valley, Washington, upstream from confluence with Palouse River and downstream from Little Goose Dam.

Figure 4.2 Rhythmites at Zillah, Washington, in the Lower Yakima River Valley.

Furthermore, scientists lately have come to believe that there were dozens of ice ages during what is called the Pleistocene. According to the uniformitarian theorists, it is possible there could have been hundreds of Lake Missoula floods prior to the last ice age (Patton and Baker, 1978b; McDonald and Busacca, 1988; Bjornstad, Fecht, and Pluhar, 2001). The question of whether there were multiple floods and ice ages before what is considered the last ice age is beyond the scope of this monograph. Each flood and ice age would have destroyed most, if not all, of the evidence for the previous events, leaving any evidence of previous ice ages and floods patchy and equivocal at best. The ice age itself will be considered in Chapter 8. We will only address the question of the number of Lake Missoula floods occurring at the peak of the last ice age since we have abundant information about that event.

BRETZ AND THE MULTIPLE FLOODS HYPOTHESIS

When J Harland Bretz was doing his landmark research in the Channeled Scabland it was commonly believed that two ice ages occurred during the Pleistocene. So, it was natural for him to postulate two floods from these two glaciations (Bretz, 1924). Bretz at first thought there was only one giant flood during the last ice age. One flood was later reinforced by the detailed work of Victor Baker (1973; 1978b,c, e).

During a field trip to the Channeled Scabland in the 1950s, Bretz shifted his opinion from one flood to many floods. He thought he saw evidence for at least seven floods, five down Grand Coulee and two affecting only the Columbia Valley west of Grand Coulee (Bretz, Smith, and Neff, 1956, p. 1045). In his last paper on the subject, he concluded that there were at least eight floods (Bretz, 1969), but it was unclear whether these floods occurred during the last ice age or during previous supposed ice ages. Bretz at that time was influenced by the revised ice age chronology of Richmond *et al.* (1965) that postulated multiple stages within each of three ice ages over most of the mountain ranges of the western United States. Even this three-ice-age chronology has since been supplanted to include dozens of ice ages (Sibrava, Bowen, and Richmond, 1986).

Was Bretz simply conforming the number of floods to the change in the number of suggested Pleistocene ice ages, or was Bretz persuaded by field data? The answer is probably a bit of both. Bretz defended his change of opinion to more than one flood by pointing out details of Beverly bar across Lower Crab Creek at its junction with the Columbia River and the multiple channels at different altitudes on the expansion bar in Quincy Basin (Bretz, 1969). He attempted to show a time sequence for certain bars, channels, and scablands that he surmised were separate floods. Later workers, however, admitted that many of these features could be explained by one large flood that waned with time (Baker, 1978e, p. 35):

Figure 4.3 Rhythmites from near Granger, Washington, in the Lower Yakima River Valley.

Figure 4.4 Rhythmites from Starbuck in the Tucannon River Valley, about 4 miles (6 kilometers) upstream from its confluence with the Snake River.

> Yet in the 1970s it appeared that such relations *could* be accomplished by just one colossal flood and its wane, there being no compelling *independent* evidence of several separate great floods (Waitt, 1994, p. k5).

Although researchers still believed in one colossal flood in the 1960s and 1970s, the idea of multiple floods was in the air at the time, thanks to Bretz himself.

Figure 4.5 Picture of three rhythmite units A to C in Burlingame Canyon. Note the "rodent burrow" (arrow) in the ripple drift cross lamination, Unit B, which isn't well developed in these rhythmites.

Figure 4.6 Burlingame Canyon, Walla Walla Valley, Washington. Note man for scale.

FORTY FLOODS BASED ON BURLINGAME CANYON

The best area for examining the possibility of multiple Lake Missoula floods is in protected tributary valleys along the edge of the flood path. These valleys conceivably would have received sediments from the flood or floods when floodwater backed up into these valleys from the temporary lake ponded north of Wallula Gap (see Figure 1.6). These slackwater sediments would not have been completely eroded as the water drained out at the end of the flood or floods. There are many tributary valleys along the main flood pathway where slackwater deposits would have accumulated. Some of these valleys are choked with giant gravel bars near their entrances with foreset beds in the bars commonly dipping *upstream*. The bars on the Snake River (Figure 4.1) dip upstream east of the confluence with the Palouse River. Other valleys, such as the Yakima, Tucannon, and Walla Walla River Valleys, show deposits from slower currents. The Tucannon River Valley joins the Snake River about 80 miles upstream from the confluence of the Snake and Columbia Rivers, which is at Pasco, Washington. The slackwater deposits sometimes are 25 to over 100 feet (8 to 30 meters) deep in the valley bottoms forming rhythmites (Figures 4.2 to 4.4). They were named the Touchet Beds by Flint (1938) after the type area near the small town of Touchet in the Walla Walla Valley. It is from these slackwater deposits in these valleys, especially the Walla Walla River Valley, where Richard Waitt discovered what he believes is evidence for 40 Lake Missoula floods.

Each rhythmite is composed of three units from bottom to top: unit A, plane-bedded coarse sand or cobbles; unit B, fining-upward ripple drift cross-laminated sand; and unit C, massive silt (Figure 4.5) (Baker and Bunker, 1985, p. 13; Shaw *et al.*, 1999, p. 606). The rhythmites are extensively eroded, especially at the valley mouths, so that only erosional remnants remain. They decrease in grain size and thickness upvalley (Baker and Bunker, 1985, p. 13), just what one expects for upvalley flow. There are also minor downvalley flow features within the rhythmites (Waitt, 1980, p. 672). A typical rhythmite is similar to a turbidite with the basal unit A, showing foreset beds dipping upvalley, probably laid down by traction. The silt of unit C is well sorted, mostly quartz grains that are similar to the Palouse silt that was eroded

Figure 4.7 Another view of Burlingame Canyon from inside the canyon looking south.

Figure 4.8 Ash layer in rhythmites 1 mile (1.6 kilometers) north of Mabton, Washington, in the Lower Yakima River Valley.

during the flood. The rhythmites represent a fairly high-energy flow in the basal portion followed by a low energy deposit in the upper portion. Exotic rocks, such as granite, are also found within the rhythmites (see Figure 2.8). It would be very difficult for these rocks to have rolled into place; the most logical mechanism is iceberg rafting during the Lake Missoula flood or floods.

Bretz at first was unsure how these slackwater deposits fit into his Spokane flood hypothesis (Bretz, Smith, and Neff, 1956). He later recognized that the debris was carried into the valleys by turbulent upvalley currents of the Spokane flood with little supposed erosion in the downcurrent flow (Bretz, 1969). Bretz was amazed that sand and silt were carried as far away as the lower Clearwater River drainage near Lewiston, Idaho. This is about 100 miles (160 kilometers) up the Snake River from its junction with the Palouse River from where the floodwater first entered the valley by spilling over a ridge to the north (see Chapter 9). He also hypothesized that there are too many rhythmites for one flood, but he thought the rhythmites were too thin to assign each one to a separate flood influx (Bretz, 1969, p. 533). Thus, Bretz introduced confusion over the interpretation of these rhythmites.

The best exposure of the rhythmites is at Burlingame Canyon, in the Walla Walla Valley, 8 miles (13 kilometers) upvalley from the valley narrows and 2.5 miles (4 kilometers) south of Lowden, Washington (Waitt, 1980. p. 655-659). (See Figure 4.11 for the location of Burlingame Canyon in the Walla Walla Valley.) Burlingame Canyon is a narrow 120 feet (35 meters) deep and 1,500 feet (450 meters) long erosional slice with vertical walls sometimes referred to as the "Little Grand Canyon." (Because of the danger, landowner permission with a notarized "hold harmless" agreement is required to view Burlingame Canyon.) There are 39 graded rhythmites exposed in Burlingame Canyon (Figure 4.6 and 4.7). The lowest rhythmites are not exposed, so there actually are more than 39 rhythmites. The rhythmite couplets are 3 to 6 feet (1 to 2 meters) thick near the bottom and thin with height to about 9 inches (23 centimeters) at the top (O'Connor and Waitt, 1995b, p. 101). Each rhythmite is strikingly similar, as emphasized by Waitt (1980, p. 657,658): "The regularity and similarity of the 39 numbered cycles are far more impressive than their variations."

Of particular interest is a white volcanic ash layer an inch or two (up to 5 centimeters) thick on the 28th rhythmite, counting from the bottom, or the 11th from the top. This ash layer can be found within the rhythmites in the other valleys of south central Washington and is very likely from an eruption of Mount St Helens (Figure 4.8). The ash is dated as 13,000 years BP (before present) (Mullineaux *et al.*, 1978). Researchers such as Baker thought the rhythmites could have been deposited in one flood. If that were the case, how could a fairly pure layer of volcanic ash be laid down in the midst of a very muddy flood that lasted about a week? This ash layer essentially gave birth to the modern multiple flood hypothesis by Richard Waitt and others (Parfit, 1995). O'Connor and Waitt (1995a, p. 56) describe how this volcanic ash became the "Rosetta stone," the key piece of information, for concluding that each rhythmite represented a separate large Lake Missoula flood:

> When rhythmic stacks of sand-silt beds in the Walla Walla valley were revisited in late 1977, the Mount St. Helens "set-S" ash couplet was found within the sequence, atop one particular bed that was not substantially different from any other bed in the section. Yet how could this be, if all beds were deposited by just one great flood?...Burlingame Canyon thus became the "Rosetta stone" for deciphering similar beds all over the region.

The reasoning went like this: since the top of each rhythmite looks the same, and the ash indicates subaerial deposition, then the top of each rhythmite must represent subaerial exposure, and hence each rhythmite indicates a separate flood separated by many years. As a result and since the bottom rhythmites were not exposed, forty floods were postulated using the data from Burlingame Canyon (Waitt, 1980, 1984, 1985; Waitt and Thorson, 1983).

The multiple flood hypothesis caught on quickly. From then on, it was easy to see further evidence for more than one flood at Burlingame Canyon and at other locations. Besides the ash layer, Richard Waitt (1980) lists the following main evidences from the Burlingame Canyon rhythmites: 1) loess or wind-blown silt and slope wash material at the top of each rhythmite, 2) rodent burrows, and 3) channels with near-vertical walls near the bottom of the sequence. He argues that loess deposition indicates a dry environment between floods. During the 40 years or so between each flood, rodents are expected to have colonized the area, burrowing into the rhythmites. Figure 4.9 shows typical features in Burlingame Canyon interpreted as a rodent borrow (see also Figure 4.5). In regard to the channel walls in the lower rhythmites, he believes these could only be cut by rainwater or small floods during a fairly long interval between major Lake Missoula floods.

Waitt also correlated these rhythmites with other rhythmites from as far away as the Missoula Valley of Montana (see Figure 5.6) and the Willamette Valley south of Portland, Oregon (Glenn, 1965) (see Figure 5.7). The rhythmites from

Figure 4.9 "Rodent burrows" from Burlingame Canyon, Washington, in the ripple drift cross lamination, Unit B.

Figure 4.10 Flood/rhythmite sequences in the Sanpoil Valley, Washington.

the Missoula Valley are a product of glacial Lake Missoula. They are composed of thick silt beds separated by silt/clay rhythmites that are interpreted as varves. The silt layers vary from 2 to 63 inches (5 to 160 centimeters) thick with an average of 12 inches (31centimeters) thick (Chambers, 1984, p. 30). They are separated by 9 to 58 "varves" with an average of 27 varves. These are the presumed number of years separating each silt layer, except that the tops of many silt/varve sequences are thought to have been eroded and weathered. So, the varve years are actually considered to be minimum estimates. Waitt and others interpret each silt unit and the overlying "varves" as representing *one* refilling of glacial Lake Missoula (Alt and Chambers, 1970; Chambers, 1971; Chambers, 1984; Alt, 1987). There are about 36 sequences of massive silt and "varves" in the Ninemile location, but the bottom of the sequence is not exposed. These sequences are also considered to be a match for the number of shorelines on the hills above the city of Missoula, supporting the connection between one rhythmite and one shoreline (Alt and Chambers, 1970).

ONE HUNDRED FLOODS?

Since the bottoms of the rhythmites at Burlingame Canyon and at Ninemile are not exposed, forty floods is considered a minimum. So, other sections at the periphery of the flood path were analyzed to ascertain what is considered the real number of floods during the last ice age.

Thus, the number of floods soon mushroomed from forty to about ninety (Atwater, 1984, 1986). This extrapolation was based on borehole sequences of rhythmites, separated by non-rhythmites, formed in the Sanpoil River Valley, a northern tributary of the Columbia River northeast of Grand Coulee Dam (Figure 4.10). The non-rhythmites, mostly graded sand layers, were interpreted as Lake Missoula flood deposits. The rhythmites are silt/clay couplets and assumed to be varves, each couplet was believed to be a one-year depositional unit. Each sequence of varves was believed to represent the time between Lake Missoula floods. There are generally 30 to 50 "varves" between each postulated flood deposit, suggesting about 40 years between each flood. The cores revealed ninety flood/varve sequences, suggesting 90 Lake Missoula floods, which were rounded off to one hundred. The total time for the one hundred floods, all from the last ice age, is at least 3000 years (Waitt, 1994, p. k-8).

Table 4.1 presents a summary of the main evidence for multiple floods as deduced by Waitt, Atwater, and others.

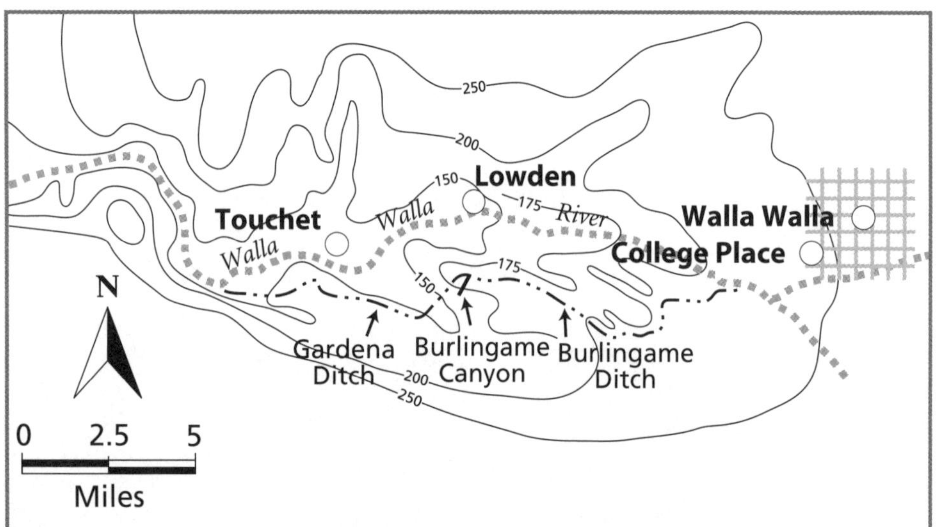

Figure 4.11 Map of Walla Walla Valley showing the location of Burlingame Canyon, about 2.5 miles (4 kilometers) south of Lowden, Washington, and the general east-west centerline ridge separated by subvalleys to the north and south. Irrigation ditch represented by a dash-dotted line. Contours every 50 meters except 175-meter intermediate counter drawn around Burlingame irrigation ditch (redrawn from maps of the U.S. Geological Survey).

Figure 4.12 View south into the southern subvalley of the Walla Walla Valley from the east-west ridge at Burlingame Canyon.

Evidence for multiple Lake Missoula floods
Ash layer at top of one rhythmite
Rodent burrows in the rhythmites
Channels cut in lower rhythmites
Correlation of rhythmites from many locations
90 flood/varve layers in Sanpoil Valley

Table 4.1 Summary of the main arguments for multiple Lake Missoula floods.

FULL CIRCLE BACK TO BRETZ'S ONE FLOOD?

The multiple floods hypothesis became widely accepted by scientists in the 1980s based especially on the idea of one slackwater rhythmite represents one flood. However, other evidence did not fit as neatly with so many floods. Gary Smith (1993) agreed with Waitt that each rhythmite in the Walla Walla and Yakima Valleys likely represented a separate flood, but he also stated that one flood is capable of laying down *more than one* rhythmite. Despite the volcanic ash layer, Smith opened up the possibility that the sequence of rhythmites in slackwater tributaries had been misinterpreted.

Soon afterwards, John Shaw and seven colleagues from the University of Alberta at Edmonton reanalyzed the Channeled Scabland from a fresh perspective and concluded that there was *only one colossal Lake Missoula flood* (Shaw *et al.*, 1999; Oard, 2000a). They further proposed that another independent flood flowed under the ice in southeast British Columbia and the Okanogan River Valley and exited south into the Columbia River. This subglacial flood happened to be concurrent with the Lake Missoula flood and served only to intensify the flooding of the western scabland area. Shaw and colleagues' deduction remains controversial, and their evidence seems to have been ignored or rejected (Jaffee and Spencer, 2000). If Shaw is correct, the number of Lake Missoula floods will have gone full circle back to Bretz's original position of one flood in the 1920s! At least, the study opens up the possibility that the number of floods is based on uncertain interpretations of complex data.

GEOMORPHOLOGY OF THE WALLA WALLA VALLEY INDICATES ONE FLOOD

Since the early 1990s I have been pondering the question of the number of floods. One hundred floods for a period over 3000 years at the peak of the ice age is too much time for the Scriptural time scale. I am convinced that the peak of the ice age probably lasted only 100 years (Oard, 1990).

Figure 4.13 View north into the northern subvalley of the Walla Walla Valley from the east-west ridge at Burlingame Canyon.

Figure 4.14. Irrigation ditch along flat east-west ridge near top of Burlingame Canyon (view ENE).

However, I was open to the possibility of several floods at glacial maximum. With the question of the number of floods in mind, I examined Burlingame Canyon a number of times over the years. From the well-publicized pictures of the rhythmites and recognizing the violent nature of flooding, I at first expected the canyon to be in a protected tributary valley at the edge of the larger Walla Walla Valley. This is because the initial flood current from each flood roaring up the Walla Walla Valley should have eroded most if not all of the previously laid *soft* sediment from previous flooding. The down valley currents as the lake in the Pasco Basin emptied through Wallula Gap were expected to be weaker but also should erode the rhythmites. Only in some protected pocket would a nice even series of rhythmites, as seen at Burlingame Canyon, collect showing no or very few signs of erosion taking place between floods. That is why I was convinced that a protected spot for the rhythmites would be necessary.

I was surprised to discover that Burlingame Canyon was *not* in some side valley but in the *middle* of the Walla Walla Valley on a ridge where it would have been on the front line of each flood assault (Figures 4.11). Notice in Figure 4.11 that the valley is narrow at its western entrance, but widens eastward. Burlingame Canyon is a south trending coulee cut perpendicular to a long generally east-west ridge. The ridge is about 1 mile (2 kilometers) wide and 125 feet (40 meters) high while the lower sections of the subvalleys north and south are each about 4 miles (6 kilometers) wide (Figures 4.12 and 4.13). The ridge extends westward from just southwest of the city of Walla Walla for about 9 miles (14 kilometers), reaching an aqueduct (dashed line) that takes the water down to the Gardena irrigation ditch. The ridge pinches out about a mile (1.6 kilometers) west of Burlingame Canyon. The top of this ridge is so flat that the Burlingame irrigation ditch runs along the length of it (Figure 4.14), represented by a dash-dotted line in Figure 4.11. That in itself amazed me since about one hundred supposed floods, many several hundred feet (over 100 meters) deep, were said to have swept up and down the valley. The water for at least the largest of the floods reached about 600 feet (180 meters) above the top of the ridge at flood maximum. I was surprised that during all the many postulated floods, *no* erosion occurred on top of this ridge while each rhythmite was evenly deposited–flood after flood. This ridge contrasts with the remainder of the valley to the north and south in which the rhythmites have been almost totally removed by erosion, even leaving lens-shaped erosional remnants (Figure 4.15). Thus, approximately 95% of the volume of rhythmites has

Figure 4.15 Streamlined rhythmite hill near Lowden, Walla Walla Valley, Washington. Long axis of hill is east-west parallel to current of the Lake Missoula flood.

Figure 4.16 Side ditch from main irrigation ditch just above Burlingame Canyon. From this ditch Burlingame Canyon formed in only six days from rerouted irrigation water (a small local flood).

been eroded out of the Walla Walla Valley. The question I found myself asking is why was there little or no sign of erosion in the very middle of the valley? This east-west ridge should have been repeatedly eroded. It seemed very surprising to me that 40 to 100 floods would have failed to erode even part of the ridge. The erosion that did take place was always concentrated in the wide subvalleys to the north and south of the centerline ridge (Figure 4.12 and 4.13).

Figure 4.17 Back side of Burlingame Canyon (view north).

Could the ridge be made of harder sediment that is more resistant to erosion? I found the rhythmites are actually very soft and easily erodible. This fact was amply demonstrated by the formation of Burlingame Canyon itself. It was excavated in only 6 days when water from the irrigation ditch that runs the length of the ridge was diverted down a side channel (Carson, 1935; Morris, 2002). The side channel was only 10 feet (3 meters) deep and 6 feet (2 meters) wide at the time (Figure 4.16) and dropped south into nearby Pine Creek. In March, 1926, persistent, strong winds piled tumble weeds into the irrigation ditch near the concrete constriction at the diversion canal, causing the water to back up. The wind blew steadily for six days and all 80ft^3/sec (2.6 m^3/sec) flow within the irrigation ditch was rerouted down the diversion canal because the Ditch Riders feared it would flood the agricultural fields. This diversion canal was not lined, so the ditch soon became a gully, which then became a gulch, and finally a canyon. The water eroded 2,000 tons (1.8 million kilograms) of rhythmites an hour and removed about 4.7 million ft^3 (133,000 m^3) of earth–all in 6 days from a flow of only 80 ft^3/sec (2.6 m^3/sec). Of course, I was left with the question of how this ridge could have possibly withstood more than one Lake Missoula flood when a very local flood so easily gouged out the sediments.

Assuming that somehow the centerline ridge miraculously avoided erosion during multiple floods, the lack of cut and fill structures at the edge of the ridge reinforced my conclusion that there was only one flood. If more than one flood impacted the Walla Walla Valley, repeated signs of erosion and deposition should be evident along the north and south edges of the ridge as each flood eroded the edge of the ridge and laid down a rhythmite. The next flood would drape a rhythmite over the eroded area creating an angular unconformity with the stack of rhythmites already deposited. This scenario would repeat itself dozens of times. In other words, the north and south edges of the ridge should show many angular unconformities and cut and fill structures from numerous erosional and depositional events. What we see are sharply truncated horizontal rhythmites at the top along the edge of the ridge with no evidence for cut and fill structures nor angular unconformities (Figure 4.17, see also Figure 4.12).

The sequence of rhythmites at Burlingame Canyon and the centerline ridge have the geomorphology, or surface features, of one large flood depositing rhythmites across the entire valley during the upvalley flow of the flood. This is followed by the almost complete removal of the rhythmites during downvalley flow along the sides of the valley as the temporary lake in the Pasco Basin drained. The ridge must have acted like a "Goat Island" which diverts water around it at Niagara Falls and is not eroded. For some reason, this double subvalley structure is typical of the Lake Missoula flood in the Channeled Scabland, as shown by the formation of other "Goat Islands" associated with the "dry" waterfalls (see Chapter 2). There does not appear to be any other way to explain the geomorphology of the rhythmites in the Walla Walla Valley other than by *one* colossal flood.

THE BURLINGAME CANYON RHYTHMITES

In addition to the geomorphology of the central ridge and its margins, the character of the rhythmites within Burlingame Canyon, is not conducive to the one rhythmite per flood hypothesis. As already stated, there is little, if any, erosion between the rhythmites, which is very unlikely with multiple catastrophic floods. At the very least, the first rush of water up the valley from each flood should be erosive enough to dissect, if not totally obliterate, these soft rhythmites. The decrease in thickness upward of rhythmites is another indication of only one flood. This feature would occur in a flood that wanes with time (Baker, 1978b, p. 65). This seems to be the most straightforward interpretation of these thinning upward rhythmites. However, Waitt and colleagues explain

Figure 4.18 Clastic dikes in Burlingame Canyon.

Figure 4.19 Close up of clastic dike from near Granger, Lower Yakima River Valley, showing vertical alternating coarse and fine layers.

the upward thinning rhythmites as indicative of smaller successive floods that occurred as glacial Lake Missoula filled to a smaller and smaller volume as the height of the ice sheet dwindled. This is a reasonable subsidiary hypothesis for the multiple floods model.

A third feature of the canyon supporting the one flood hypothesis is the ubiquitous clastic dikes forming a polygonal pattern in the rhythmite sequence (Figure 4.18). These dikes also occur in the rhythmites from other valleys of south central Washington. The dikes are 1 mm to a few meters thick with variable dips and strikes, but most of the large ones are nearly vertical, cutting through the *whole* sequence. These dikes are unique in that they *repeat* the rhythmite sequence of sand and silt with even a few silt rip-up clasts (Figure 4.19). One would not expect clastic dikes slicing through the whole sequence, if a separate flood deposited each rhythmite. If each rhythmite was deposited separately, then each rhythmite should have its own set of dikes, generally separated from other rhythmites with most of the dikes ending at the top of each rhythmite. Clastic dikes appear to be evidence for rapid deposition of the whole rhythmite sequence during one flood (Rigby, 1982, p. 45).

The origin of the clastic dikes in Burlingame Canyon and other slackwater rhythmites has been controversial. Were these unique dikes filled from upward pressure (from below) or were they formed as a result of cracking and filling (from above) during desiccation (Jenkins, 1925; Lupher, 1944; Carson, McKhann, and Pizey, 1978; Carson and Pogue, 1996)? Many possible explanations have been proposed. Cooley, Pidduck, and Pogue (1996) from the Geology Department at Whitman College in Walla Walla think they were formed by a downward injection of sediment from an earthquake or a series of earthquakes. Then, where did this infilling sediment originate, and how could such a mechanism produce the unique vertical banding in the dikes? It seems more likely the dikes formed from below as a result of upward pressure. Shaw *et al.* (1999, p. 607), Bjornstad (1980, p. 75), and others, believe the clastic dikes formed quickly upward from below. They think the abundant water in the rhythmites during one flood was under pressure when the lake was at its deepest. Then, when the water rapidly drained through Wallula Gap, the pressure above was reduced in a few days, while the pressure in the groundwater within the sediments remained high. This caused an explosive release of water through the sediments forming the clastic dikes. It is possible earthquakes at this time were involved as well (see Chapter 5). This hypothesis seems likely.

THE ENIGMATIC WALLULA GAP RHYTHMITES

About two-dozen rhythmites are exposed in a railway cut one mile (1.6 kilometers) northeast of Wallula Gap (Figure 4.20), strongly indicating there was only one flood (Bjornstad, 1980, p. 75). There probably are more rhythmites not exposed below the sequence. Bjornstad, Fecht, and Tallman (1991, pp. 237,238) state the significant of these deposits:

> Particularly damaging to the hypothesis that the rhythmites represent "periodic, colossal jökulhlaups" (Waitt, 1985) is the fact that these low-energy deposits commonly occur at sites that would have experienced high velocities and stream power during a cataclysmic flood. For example, a sequence of two dozen or more rhythmites, deposited under low-energy conditions, is preserved at the mouth of Wallula Gap, an environment that would have been subject to phenomenal flow-velocity conditions at the onset of any cataclysmic flood (Bjornstad, 1980).

Other scientists believe these rhythmites would not have remained if there were more than one flood (Baker, 1989b; Baker *et al*, 1991).

O'Connor and Waitt (1995b, pp. 99,104), although admitting that the rhythmites seem incongruous, dismiss this evidence as "...deductive speculation whose premises fail to include elemental field relations." This comment is perplexing. They point out that ripple marks on the rhythmites show a northward current so that the rhythmites were laid down in an eddy. Deposition in an eddy is a reasonable deduction, but it still does not support the multiple flood hypothesis in which each flood would arrive with a catastrophic rush of water that would destroy such delicate layers. No sequence of two-dozen rhythmites from two-dozen floods could build up and survive in one spot subject to high-energy flow.

This sequence of rhythmites provides powerful support for only one flood. It is true that in one flood the initial rush of water would have eroded any soft sediments, but the rhythmites would have been deposited later and probably over a larger area as the water rose in the ponded lake. As the temporary lake drained, eddies would have formed at the edge of the main flow. Soft sediments would be eroded close to the gap and in fast water and then would have been preserved in an eddy, leaving behind an erosional remnant.

Very strong evidence favors one colossal Lake Missoula flood. Table 4.2 provides a summary of the evidence for only one flood.

Evidence against multiple Lake Missoula floods
No erosion of east-west ridge in middle of Walla Walla Valley
Lack of unconformities at edge of east-west ridge
No or little erosion between rhythmites at Burlingame Canyon
Fining upward rhythmites at Burlingame Canyon
Some clastic dikes cut through all rhythmites
Two dozen rhythmites near Wallula Gap

Table 4.2 Evidence against more than one colossal Lake Missoula flood

MINOR LAKE MISSOULA FLOODS?

Just because there was one monstrous Lake Missoula flood does not rule out the possibility of other minor floods. It is likely that after one catastrophic dumping of glacial Lake Missoula, the ice in the Purcell Valley of northern Idaho surged southward to fill the gap and form another ice dam and glacial Lake Missoula. The amount of water backed up by the dam would have depended on the height of subsequent ice dams, which is based on the characteristics of the ice age. Within the uniformitarian system, glacial maximum during the last ice age would have lasted probably several thousand years. Dozens of Lake Missoula floods is a reasonable deduction from this model. However, if the ice age was short and glacial maximum lasted on the order of a hundred years, only one colossal flood is a reasonable deduction. In this model of a rapid ice age (Oard, 1990), the lake following the major flood would have been much smaller for two reasons: 1) the ice would have spread its volume from north of the

Figure 4.20 Rhythmites near Wallula Gap.

dam breach over a significantly large area when it surged southward, and 2) the ice dam would be melting rapidly not building up. Although melting would cause the lake to fill more rapidly, it would also have kept the ice dam from growing. There is no doubt that one colossal Lake Missoula flood supports the rapid ice age model and not the uniformitarian model. Furthermore, at least one more partial filling of glacial Lake Missoula is likely under the rapid ice age scenario. Whether it caused another flood of lesser intensity is another story, since a shallower lake could have drained slowly and not catastrophically through its ice barrier.

There is evidence that a smaller glacial Lake Missoula formed after the main flood. This comes from the Camas Prairie gravel ripples. These were described in Chapter 2. Lister (1981) discovered lake sediments, "varves" as he called them, *deposited on top of* the large ripple marks. The so-called varves are about 6 feet (2 meters) thick in the lower, southern part of the valley and thin northward and upward to about 1 inch (2.5 centimeters). Moreover, he discovered faint shorelines along the faces of two of the deltaic bars exiting from the northern passes. These deltas were formed from the catastrophic draining of the larger glacial Lake Missoula. So these shorelines were etched in these gravel bars *after* the main event. The highest shoreline is 3360 feet (1025 meters) ASL, which is about 800 feet (245 meters) lower than the maximum lake level. The volume of this lake would have been about 135 mi^3 (550 km^2), one quarter the maximum volume (Craig, 1987, p. 319). Lister (1981, p. 45) concludes:

> What this study has proven is that there was a minimum of two lake stands, separated by one major flood. There could have been several, earlier floods that preceded the one recorded in Camas basin, but all must have been lesser because there is no evidence of reactivation surfaces within the [gravel bar] bedforms. Likewise, there could have been numerous smaller floods post dating this one.

The evidence from Camas Prairie points to two lake stands, a very large lake and a small lake. Because the gravel ripples show no sign of reactivation surfaces, it points to only one major flood as the larger lake catastrophically drained. There is no real evidence of lesser floods before the big one, since the major flood would have destroyed evidence of previous floods. All we know is that the lake likely refilled at least once up to 3360 feet (1025 meters) after the one major flood. We do not have any evidence from this one locality of lesser floods after the major flood. This evidence must be ascertained from eastern Washington.

After this minor refilling of the former glacial Lake Missoula, several events could have transpired, depending on a number of variables. The lake may have discharged slowly through its ice dam, causing high water for many years on the Columbia River. It could also have completely broken through its ice dam, initiating another Lake Missoula flood but with one quarter less water. This flood would have spread west through Spokane and into glacial Lake Columbia. From there it would have spread south across the western scablands through Grand Coulee, since it is known that glacial Lake Columbia survived the monstrous Lake Missoula flood and any minor floods afterwards (Atwater, 1987). This minor flood would have missed the Cheney-Palouse and Telford-Upper Crab Creek scabland tracts because their north to northeast entrances start too high above the Spokane and Columbia Rivers.

It is even conceivable that the above sequence of events could repeat itself a few more times during glacial melting, each time with a smaller ice dam and less water. Maybe these smaller floods are what Bretz noticed in the details of the western Channeled Scabland. For instance, Beverly bar is 100 feet (30 meters) high and 1.5 miles (3 kilometers) long, and it blocks the Lower Crab Creek, a main drainage valley from Quincy Basin through the Drumheller Channels and into the Columbia River. This bar implies that a flood higher than 100 feet (30 meters) swept down the Columbia River *after* the main event (Bretz, 1969, p. 525). This floodwater probably would have spread out in the expansive Pasco Basin and been significantly shallower that the main Lake Missoula flood. It would have been a significant event, of course, but not nearly as dramatic as the main Lake Missoula flood. On the other hand Beverly bar could be a waning remnant of the Shaw and colleagues (1999) subglacial flood down the Columbia River from British Columbia. A third possibility is that the bar was formed in the major Lake Missoula flood after water stopped flowing down Lower Crab Creek. Regardless, the evidence for minor floods in the western scabland is uncertain.

Chapter 5

How Can Evidence for Multiple Floods Be Explained?

The evidence strongly points to one Lake Missoula flood. Then how can the data used to support multiple floods be explained within a one-flood model? How, especially, is the band of volcanic ash in the rhythmites, the "Rosetta Stone," to be explained? What about Waitt's other evidence from Burlingame Canyon? Atwater has provided evidence for 90 floods over a 3000-year period in the Sanpoil Valley based on graded sand layers separated by varves. How is the Sanpoil Valley data to be explained? Are there really 3000 varves, each representing a year? Do the rhythmites from western Montana correlate to rhythmites in eastern Washington and the Willamette Valley? Does each shoreline in western Montana represent another lake filling? These "evidences" for multiple floods will be addressed in this chapter.

SLACKWATER RHYTHMITES FORMED BY SURGES

The most important question, which led to widespread belief in multiple Lake Missoula floods is: how a fairly-pure volcanic ash band can be deposited in slackwater sediments during a flood that lasted only a week? This ash was considered to be the "Rosetta Stone" for Waitt to postulate 40 floods from Burlingame Canyon. But, before I address this issue, I need to explain how the stacks of rhythmites were likely formed in the first place.

I believe the explanation put forth by Bretz and Baker is reasonable and correct. They believed that surges or pulses of water surging up the slackwater tributaries formed the rhythmites. Bretz (1929b, p. 539), himself, pointed to this possibility. Baker expanded on it, considering the rhythmites to be a type of *turbidite*:

> Any disturbance in the water surface of the main scabland channels was propagated up these tributaries in the form of transient surges (water-surface waves). Such surges would bring into the tributary valleys a mixture of main channel flood sediments in the form of a density flow or turbidity current (Baker, 1973, p. 46).

Turbidites can be deposited within minutes or hours, so each rhythmite easily could represent an upvalley surge during a single flood.

Several mechanisms have been proposed that would cause these surges (Baker, 1978b, p. 66; Carson, McKhann, and Pizey, 1978; Bjornstad, 1980, pp. 60-62). First, a variable discharge from glacial Lake Missoula would have occurred as the floodwater wound its way through the torturous valleys and narrows of western Montana. Second, there would have been convergent and divergent water flows in the scabland because of the numerous and varied channels, splitting and recombining downstream. Convergent flow would result in deeper water and a surge wave. The higher water surface of the convergent flow would have caused it to flow faster and pick up more sediment. A divergent flow would result in slower currents. Third, landslides, ice blocks, or gravel bars would have temporarily interrupted water flow in some of the scabland channels, causing a surge. Fourth, sudden channel deepening would cause a surface wave. Fifth, the flow from the main Pasco Basin would interfere or converge with flows from tributary valleys. All of these mechanisms could cause multiple surges and rhythmite beds to form in the tributary valleys as the surges repeatedly propagated up the tributary valleys.

Waitt (1980, pp. 671-673) dismissed the significance of surges for the rhythmites. However, multiple rhythmites have been *observed* to form during surges in a single flood. For example, well-stratified slackwater beds have been deposited in tributary canyons during flooding of the Pecos and Devils Rivers in West Texas (Patton, Baker, and Kochel, 1979).

A catastrophic outburst flood (jökulhlaup) in Iceland provides a close analog to the deposition in the tributary valleys of south central Washington during the Lake Missoula flood (Russell and Knudsen, 1999). A subglacial volcanic eruption melted some of the ice under the glacier. The water flowed under the ice and burst from the edge on November 5, 1996. The peak discharge was 1.6 million ft^3/sec (45,000 m^3/sec), only 0.2% of the peak flood discharge of the Lake Missoula flood. The Icelandic jökulhlaup lasted 36 hours. During the later half of the flood, the discharge from under the ice switched outlets, so that the original outlet became a *slackwater embayment*. The embayment rapidly filled with sedimentary layers from the slackwater flooding. The flood produced planar beds dipping at a low angle. These planar beds are actually rhythmites of normally graded fine gravel and course sand. Two hundred planar rhythmites and 100 prograding rhythmites formed a section 50 feet (15 meters) thick in just

Figure 5.1 Ash band in rhythmites from western Badger Coulee, just south of Pasco, Washington. Notice that the ash forms two couplets.

17 hours. This is one rhythmite every 3 to 4 minutes! These rhythmites were observed to be deposited by repeated turbulent flow *pulses*. So, the formation in one week of 40 or more relatively thick rhythmites in tributary valleys due to repeated surges during one Lake Missoula flood is easily conceivable, especially in view of the fact that the Lake Missoula flood eroded 50 mi^3 (200 km^3) of basalt and overlying silt.

A VOLCANIC ASH BAND DEPOSITED IN ONE FLOOD?

It does appear to be an incredible coincidence that a volcano was erupting during the one week Lake Missoula flood. Both the flood and the eruption are rare events, and to have them occurring at the same time appears to be too much to ask for the one flood hypothesis. It seems more reasonable that the ash landed on a dry surface between floods as believed by Waitt and others. Many scientists were persuaded by Waitt's strongly stated logic, including Russell Bunker (1982, p. 27), who would normally have believed in only one flood for all the rhythmites, if there was no ash band. Although Bunker at first thought that the rhythmites could be formed by one flood, he later postulated two floods separated by the ash band for his study site in Badger Coulee, just south of Pasco, Washington (Figure 5.1).

When I first observed the ash band in Burlingame Canyon, as well as at many other locations since then (Figure 5.2), I noticed features that seemed contrary to Waitt's interpretation. The ash layer, although fairly pure, was of generally even thickness in most areas and often formed a couplet (Figure 5.1). I wondered that if the ash was deposited on dry ground and remained for up to 40 years before the next flood, why was it not partially to completely mixed with silt and sand? Wind, rain, and other processes should have mixed the ash within a short time. Mount St. Helens erupted in 1980 and spread a layer of ash up to a foot thick over most of the Channeled Scabland. Twenty years later there are very few distinct ash layers remaining and we have presumably less silt and sand blowing and less precipitation now than during the ice age. The only way for such an even band to be preserved, assuming the multiple flood hypothesis, was if the ash fell right before the next flood. Regardless of when the "next" flood occurred, the catastrophic onslaught of the following flood would easily erode and mix this ash. The ash layer in the rhythmites is too even over too large of an area to have been deposited between two Lake Missoula floods. In regard to the ash couplets, how could the silt or sand be deposited *between* the ash bands in a subaer-

Figure 5.2 Ash band in rhythmites along the Yakima River near Granger, Washington.

ial environment? It implies that both silt/sand and ash were deposited rapidly together, with only a slight interruption in ash deposition, possibly on the order of minutes.

It is more reasonable to conclude that the ash band was rapidly deposited and buried, forming couplets in areas, during one flood, therefore preserving its relative purity and evenness. Shaw and colleagues (1999, p. 607) after a detailed examination of the Channeled Scabland, arrived at a similar conclusion in regard to the ash layer:

> Waitt's (1980) photographs of the ash show dark silt and sand layers intercalated with the lighter ash, suggesting simultaneous deposition of the ash and suspension deposits. Massive silt in unit C certainly resemble loess [wind-blown silt], yet gradational relationships with climbing ripple cross-lamination and their regular position, with respect to aqueous deposits as a fining upward sequence, suggests aqueous deposition. We conclude that the ash was deposited from a water column subsequent to air fall. The ash was most likely falling during the formation of the previous rhythmite or rhythmites and ended up fairly pure at the top of the 28th rhythmite in Burlingame Canyon because of a pause in deposition before the next rhythmite was laid down.

Another damaging observation to Waitt's multiple flood position, based on the volcanic ash layer, comes from the Mabton site in the Yakima Valley. A thin ash lamina is also found on the top of the rhythmites *above and below* the rhythmite with the prominent ash band (O'Connor and Waitt, 1995b, p. 108). So it appears that the volcanic ash was falling into the water while at least three rhythmites were being laid down in the Lower Yakima Valley. In order for Waitt to maintain his multiple flood hypothesis, he must postulate *three* separate volcanic eruptions after three consecutive floods, which is unlikely.

It is still quite a coincidence that an ash layer from Mount St Helens was deposited near the top of a rhythmite sequence during only one flood. Waitt (1980, p. 665) exclaims: "The enclosure of even a single layer of tephra would be extraordinary in sediment that accumulated in less than one week." But what Waitt did not consider, and as a result postulated his multiple flood scenarios, is that it is possible for one giant flood to *cause* the volcanic eruption that laid down the ash. This deduction seems quite speculative at first, but there are sound observational reasons as to why this should be the case.

It has been known for many years that artificial reservoirs, which fill slowly after construction, cause earthquakes. The strongest earthquakes caused by a filling reservoir occurred at the Koyna Dam reservoir, west-central India, starting in 1962 (Gupta *et al.*, 2000). More than 150 quakes of magnitude greater than or equal to 4.0 were recorded. Several have been much stronger:

> At least 10 events with magnitude equal or above 5.0 have occurred since impounding of the Koyna Reservoir. To date, the largest (M6.3) reservoir-induced earthquake in the world occurred on December 10, 1967, near the Koyna Dam. This earthquake claimed over 200 human lives (Gupta *et al.*, 2000, p. 145).

A magnitude 6.3 quake is a strong quake, especially strong for one caused by a reservoir slowly filling up for the first time. Even stronger quakes should have been triggered during the Lake Missoula flood when 750 feet (230 meters) of water with a volume of about 300 mi^3 (1230 km^3) ponded within just a few days in the Pasco Basin of south central Washington.

New evidence demonstrates that strong quakes can trigger eruptions from volcanoes that are close to erupting (Linde and Sacks, 1998; Simpson, 1998; Hill, Pollitz, and Newhall, 2002). For instance the 1992 magnitude 7.3 Landers earthquake in Southeast California triggered a remarkably sudden and widespread increase in earthquake activity across most of the western United States. This activity included the Long Valley, California, and the Yellowstone volcanic fields. The Long Valley, a collapsed volcano 250 miles (400 kilometers) northwest of the Landers earthquake, shuddered hundreds of times a day, and sensitive instruments detected swelling of the pool of magma underneath the surface (Simpson, 1998). Fortunately, it did not erupt. The magnitude 7.1 Hector Mine quake of October 16th, 1999, in Southeast California also caused earthquakes in the Long Valley caldera (Hill, Pollitz, and Newhall, 2002). The Landers Fork quake is also believed to have caused small quakes 15 miles (24 kilometers) northwest of the Yellowstone Park caldera–780 miles (1250 kilometers) away (Hill *et al.*, 1993, p. 1620)! A 7.0 magnitude earthquake on southern Kodiak Island triggered sudden earthquake activity of smaller magnitude underneath volcanoes in the Katmai area of the Alaska Peninsula (Power *et al.*, 2001). A statistical study of very large quakes indicates that earthquakes can touch off volcanoes within 470 miles (750 kilometers):

> The geophysicists reported in the Oct. 29 NATURE that 8 of the study's 204 earthquakes of magnitude 8.0 or greater seemed to trigger same-day eruptions within 750 km [470 mi] (Simpson, 1998).

Such a statistical relationship is very significant, given the rarity of the events, indicating a connection between strong quakes and volcanic eruptions. The correlation is not even close to 100%, meaning that many strong earthquakes failed to cause eruptions. The reason that more volcanoes did not erupt is likely because they were not ready–the magma and pressure build up was not great enough at that time. Hill, Pollitz, and Newhall (2002, p. 41) conclude:

> New measurements, statistical analyses, and models support the conjecture that a large earthquake can trigger subsequent volcanic eruptions over surpris-

ingly long distance and time scales.

It has been noted by a number of researchers that any load added or subtracted from the earth's crust can initiate earthquakes and volcanism, such as the initiation and removal of ice during the ice age (Hall, 1982; Sigvaldason, Annertz, and Nilsson, 1992; Nakada and Yokose, 1992; Rampino and Self, 1993; Thorson, 1996). Sea level change during the ice age may also have been enough to trigger a volcano (McGuire *et al.*, 1997). Ice age meltwater loading or unloading is thought by some to possibly initiate distant volcanism (McGuire *et al.*, 1997, p. 473). One researcher suggested that the pressure drop caused by the removal of a 100 m deep glacial meltwater lake triggered an eruption in Iceland (Sigvaldason, Annertz, and Nilsson, 1992, p. 385).

Mount St. Helens is one of the most active volcanoes in North America (Figure 5.3). It is only 150 miles (240 kilometers) from the rapidly ponded lake in south central Washington. It stands to reason that the catastrophic increase in water would have triggered at least one very large earthquake that likely caused the eruption of Mount St. Helens. The ash from this eruption would have generally spread eastward in the prevailing westerly winds, similar to the 1980 eruption. The ash would have fallen in the temporarily ponded lake in south central Washington, forming an ash band or ash couplet in the slackwater rhythmites.

Figure 5.3 Mount St Helens, Washington (view south into the crater left by the 1980 eruption).

Figure 5.4 A "rodent burrow" in the ripple drift cross lamination, Unit B, in Burlingame Canyon, Washington.

MULTIPLE FLOOD EVIDENCE FROM BURLINGAME CANYON QUESTIONABLE

Wait also pointed to wind blown silt at the top of each rhythmite, rodent burrows, and channels in the lower rhythmites in Burlingame Canyon as further proof of the multiple floods hypotheses. How can these evidences be explained by one flood?

Waitt claimed that the very top of each rhythmite, unit C, in Burlingame Canyon and other slackwater rhythmites in various valleys is composed of loess or wind-blown silt that was laid down in a dry environment between floods. However, Unit C is composed of the *same* sediment as the Palouse silt that caps the basalt of eastern Washington. It is reasonable that the silt of Unit C resembles loess, but that does not mean that the very top of unit C was blown into place. Even Waitt and Atwater (1989, p. 50) admitted this interpretation is a stretch:

> Loess between any two beds indicates that a terrestrial, eolian environment intervened between the floods that deposited the two beds. But the loess is difficult to identify, for the waterlaid tops of most graded beds [of the rhythmites] are texturally *nearly identical* to loess, which is the *very source* of the flood laid sediment [italics and brackets mine].

So even the very top of the massive silt of unit C cannot legitimately be used as evidence of a subaerial environment.

What about those rodent burrows claimed by Waitt? They do appear to be rodent burrows (Figure 5.4, see also Figures 4.5 and 4.9), and even Shaw and colleagues (1999, p. 606) were perplexed at their origin. In the Tucannon Valley, they suggested that modern rodents made filled "burrows" because they never saw networks extending down from the top of any one rhythmite. I can understand rodents burrowing into rhythmites if they are not too thick, as in the Tucannon Valley and areas of the Yakima Valley, but I cannot envision rodents burrowing down a hundred feet (30 meters) into the rhythmites of Burlingame Canyon. When I investigated the so-called burrows, I discovered that they seem to be located predominantly within the coarser fraction of the rhythmites, in units A and B. Rodents would have to burrow through unit C to get to these lower units, so there should be numerous burrows in unit C, but few if any burrows can be seen in unit C.

It is quite possible that the "burrows" represent water escape structures formed by the release of water under pressure, probably formed at the same time as the clastic dikes. Units A and B are the most permeable layers, being composed mostly of sand. A similar argument can be made for the supposed root casts (Figure 5.5), but on a smaller scale. Bjornstad (1980, p. 39) also believes the "rodent burrows" and "root casts" most likely represent forceful injection of water under a hydroplastic state (Spencer, 1989, p. 172). In other words the "rodent burrows," as well as "root casts," could be rapid dewatering structures that occurred after only one flood, possibly due to the release of pressure immediately after the water drained. It is even conceivable that the seismic activity caused by the ponding of water behind Wallula Gap produced or contributed to the formation of the "rodent burrows," "root traces," and clastic dikes in the rhythmites of south central Washington.

Figure 5.5 "Root casts" from Lake Missoula flood rhythmites about 1 mile (1.6 kilometers) north of Mabton, Lower Yakima Valley, Washington.

Figure 5.6 Rhythmites in the Ninemile section, 30 miles west of Missoula, Montana, that are claimed to be a sequence of 35 or 36 silt layers separated by varves.

It has been claimed by Waitt that channels have been cut in the lower rhythmites in Burlingame Canyon, implying a subaerial environment. Shaw and colleagues (1999, p. 607) considered these channels to be only minor scours confined to the basal units. They pointed out that similar scours are found in other situations and often relate to local erosion caused by underflows. So, these channels are not that significant in distinguishing between a subaqueous or subaerial environment. Besides it is well known that loess can form vertical cliffs. In one Lake Missoula flood the channel cut would be filled in before the walls would cave in. If these "channels" did form in a subaerial environment, the "next" flood should have destroyed them.

Figure 5.7 Silt and silty-clay rhythmites along the Willamette River between Salem and Oregon City, Oregon. The site is a fresh exposure caused by slumping during the large flood in February, 1996.

Figure 5.8 Silt/clay rhythmites in the Sanpoil section, north central Washington, that are claimed to be varves.

CAN RHYTHMITES BE CORRELATED FROM WESTERN MONTANA TO WESTERN OREGON?

Waitt correlated the rhythmites from Burlingame Canyon with the Ninemile section near Missoula, Montana (Figure 5.6), and the Willamette Valley of western Oregon (Figure 5.7), claiming each rhythmite represents a separate flood. The rhythmites in the Willamette Valley, varying from silt to silty clay, are finer grained than those in south central Washington. Finer-grained rhythmites would be expected in the far reaches of the Lake Missoula flood because most of the coarser-grained sediments would have already been deposited. But the correlation of rhythmites over such long distances is questionable (Bunker, 1982, p. 28). Baker and Bunker (1985, p. 25) claim that any long distance correlation with no distinctive marker beds is not justifiable. Another problem is that the rhythmites deposited at the bottom of glacial Lake Missoula, as well as in glacial Lake Columbia, formed *within lakes*, while those in tributary valleys of the Lake Missoula flood formed as a result of flooding–two different mechanisms. And just as expected, the rhythmites from the Ninemile section, as well as from the Sanpoil Valley that were deposited in glacial Lake Columbia, differ from those in the slackwater embayments of eastern Washington and the Willamette Valley.

WHAT ABOUT THE NINETY FLOODS CLAIMED FROM THE SANPOIL VALLEY?

Another question that needs to be resolved is: if there was only one flood, how can the 90 Sanpoil Valley flood/rhythmite sequences be explained? Figure 4.10 shows a typical exposure of several sequences. Shaw and colleagues (1999, p. 606) discovered what seems to be fatal evidence for the idea that these sequences represent repeated Lake Missoula floods:

> He [Brian Atwater] interpreted rhythmic beds as varves and noted sand at the base of many sequences with *downvalley* paleocurrents, indicating discharge from the Sanpoil Sublobe [of the Cordilleran Ice Sheet] to the north. Basaltic clasts, which should occur in the Sanpoil arm with flooding from Glacial Lake Missoula, are *absent*. We suggest powerful flows from the north... [emphasis and brackets mine].

The lack of basalt clasts in deposits claimed to be from scores of Lake Missoula floods seems especially significant, since the northern edge of the Columbia River Basalt Group lies only several miles (about 5 kilometers) south of the Sanpoil Valley. A Lake Missoula flood had to traverse and erode the basalt before reaching the Sanpoil Valley and should have left basalt sand and rocks in the rhythmite sequence up the valley. This sequence in the Sanpoil Valley is more likely to have been caused by repeated meltwater flows emanating from the ice sheet to the north into glacial Lake Columbia. The sequence surely has little, if anything, to do with the Lake Missoula flood.

ARE SILT/CLAY RHYTHMITES VARVES?

Silt/clay rhythmites are observed in the Sanpoil section between strong meltwater flows (Figure 5.8) and in the Ninemile section separated mostly by silt (Sieja, 1959). They formed in glacial Lake Columbia and glacial Lake Missoula, respectively. Each silt/clay couplet is interpreted as a one-year varve. There are about 40 silt/clay couplets between 90 coarser-grained sediments in the Sanpoil section and about 28 silt/clay couplets between 35 or 36 coarser-grained layers in the Ninemile section (Chambers, 1984; Alt and Chambers, 1970). Counting varves in the Sanpoil section produced over 3000 years while in the Ninemile section only 1000 years were tallied. Alt (2001, p. 28) claims the missing 2000 years from the Ninemile section represent the time before the section was flooded as each lake started filling. The lake would have been 1000 feet deep in northern Idaho by the time the lake reached the Ninemile section. However, if the entire section at Ninemile represents only one filling of glacial Lake Missoula, then the silt/clay rhythmites could not be varves. If not varves at Ninemile, the silt/clay couplets likely are not varves in the Sanpoil section either. This would mean that 3000 years of time did not exist. But are these these silt/clay couplets really varves of one-year deposition? There was a time when silt/clay couplets were *automatically* assumed to be one-year varves. Anderson and Dean (1988, p. 216) state:

> There was little questioning of the assumption of annual deposition for proglacial lake sediments because seasonally regulated melting and freezing was so obvious, and early investigations...were convincing.

So, the one-year period for deposition of a silt/clay couplet is really an assumption. Quigley (1983, p. 150) agrees: "A single varve representing 1 year of deposition, consists of a couplet of summer silt and winter clay: this time framework is difficult to demonstrate however." So-called varves, especially in Scandinavia, have been used for over 100 years to establish a long chronology of deglaciation. In fact, the Scandinavian varves were used to construct the first "absolute" chronology in the early 1900s (Oard, 1992a). The chronology, however, was simply fit into the preconceived beliefs about the time sequence of deglaciation at the time. The "varve" chronology was not an independent chronology, and the "dates" have been shuffled around ever since (Oard, 1992a,b).

Modern research indicates there are several other mechanisms that can produce silt/clay couplets in *less* than one year (Oard, 1992a,b). In a study of silt/clay rhythmites in a Swiss Lake, Lambert and Hsü (1979) discovered that 300 to 360 couplets were formed in 160 years. Two couplets generally formed in a year due to turbidity currents caused by spring snowmelt or heavy rainstorms. Sometimes 5 couplets formed in a year. The number and thickness of lamina varied with the location, which strongly suggests that such sequences cannot be correlated from place to place. Such correlations from many locations, nevertheless, have been claimed by geologists for over a century in Scandinavia and elsewhere.

There are other examples of multiple couplets, similar to real varves, formed within one year. Pickrill and Irwin (1983) report that an average of *three* couplets were deposited each year in a glacier-fed lake in New Zealand. Wood (1947) discovered that varve-like couplets were deposited in a new reservoir by a muddy stream caused by rain showers. He noticed three couplets developed in just two weeks time.

Muir Glacier, Southeast Alaska, has been receding at an average of 1340 feet (410 meters) a year for many years. As it melted back, the glacier discharged great quantities of mud into Muir Inlet. Scientists cored the sediments being deposited in the inlet and discovered that the debris was forming numerous rhythmites, similar to varves, each year (Mackiewicz *et al.*, 1984; Cowan, Powell, and Smith, 1988). The couplets were piling up at a rate of 43 feet (13 meters) a year adjacent to the glacier and about 3 feet (one meter) a year far from the glacier! In the summer, high discharges and semidiurnal tides resulted in *two couplets a day* (Smith, Phillips, and Powell, 1990)! Muir Inlet is a good example of the environment of rapid melting and sedimentation that occurred during deglaciation of the ice age. Therefore, multiple rhythmites are expected each year in a lake that lies adjacent to a melting ice sheet.

The "varves" in the Sanpoil and Ninemile sections were deposited in lakes adjacent to glaciers. The Muir Inlet model, minus the tides, would represent the depositional environment of these sequences. Multiple silt/clay rhythmites would be expected each year.

Within the one Lake Missoula flood model, the sequences at Sanpoil and Ninemile likely represent a spring or early summer coarse layer due to melting of ice, followed by a gradual settling of silt and clay in a muddy lake from late summer through the winter. Each sequence of silt or sand followed by silt/clay couplets likely would represent one year of deposition. Thus, the Sanpoil sequence would represent about 90 years of deposition in glacial Lake Columbia, while the Ninemile section would represent about 35 or 36 years of deposition in a fairly deep lake, since it takes relatively deep water for silt/clay couplets to form.

Further evidence that the whole sequence at Ninemile represents only one major filling of glacial Lake Missoula was discovered by Shaw *et al.* (1999, p. 605). They noticed that each thick silt layer between silt/clay couplets was overlain by a thick layer of clay. The clay would represent the settling of the finer particles after the main spring or early summer melt pulse. Clay can be deposited rapidly due to coagulation of particles. Such a thick clay layer would make no sense if each large-scale rhythmite represents a refilling of glacial Lake Missoula.

Furthermore, there are little or no signs of erosion between each large-scale rhythmite. If each large-scale rhythmite represents a complete filling and breaching of glacial Lake Missoula, there should be abundant evidence of erosion within such a sequence by two mechanisms: 1) from the water rushing out of the valley during each dam breach and 2) as the lake shoreline progressed back across the section during each refilling. The edge of the whole rhythmite sequence is evenly truncated (Figure 5.9), indicating the main erosive event, which eroded a majority of the rhythmites from the valley, occurred only after they were *all* deposited. If for some reason this isolated erosional remnant remained after each flood, there should be numerous cut and fill structures and unconformities, at least at the edge of the sequence, similar to the situation along the centerline ridge in the Walla Walla Valley (see Chapter 4).

HOW WERE THE LAKE MISSOULA SHORELINES FORMED?

Waitt, Chambers, Alt, and other advocates of multiple floods have claimed that each shoreline of glacial Lake Missoula represents one complete filling and emptying of the lake. Their main evidence seems to be the similarity of the number of rhythmites at Ninemile and the number of shorelines on the hills above Missoula (Weber, 1972, p. 24). However, scientists now postulate about 100 floods, just for the "last" ice age. Since they believe in dozens of ice ages, there would have been maybe a thousand fillings of glacial Lake Missoula, each forming a shoreline or multiple shorelines. So, the forty or so shorelines observed on the hills above Missoula no longer match the presumed number of lake fillings. If there was only one flood, as developed in this book, how would these shorelines be interpreted within just one filling of glacial Lake Missoula?

Figure 5.9 Truncated rhythmites at Ninemile section, 30 miles west of Missoula, Montana. This picture is a westward extension of Figure 5.6, showing the even truncation of the rhythmites without cut and fill structures or unconformities.

Figure 5.10 Lake Missoula shorelines on Mount Jumbo viewed parallel with the shorelines. Notice slight undulation in the middle foreground.

Before providing an interpretation of the shorelines, it is useful to examine the specific properties of these shorelines. The shorelines are faint but generally uniform and regularly spaced (see Figures 3.2 to 3.4). They are so faint that they are hardly noticeable while walking over them. Viewed parallel to the sequence, each shoreline forms only a very shallow trough-ridge pattern (Figure 5.10). Several scientists have noted that the shorelines are not only faint, but also they are uniform (Bretz, 1959, p. 53). Bretz commented:

Missoula's numerous shorelines are so faint and close set that there is little hope of determining the total number or of correlating from valley to valley. No long-continued pauses in the changing lake level are on record (Bretz, 1969, p. 533).

Note that Bretz believed the shorelines did not represent long pauses. The highest shoreline also is no more marked than any other. The shorelines are uniform likely because they represent little erosion by wave action or weathering after the shorelines formed (Pardee, 1910, p. 380; Weber, 1972, p. 24). Even the shorelines etched into soft rocks are modified very little by subsequent mass wasting (Weber, 1972, p. 25).

While investigating the shorelines on Mount Jumbo, I estimated a fairly regular vertical spacing of about 25 to 40 feet. Weber (1972, p. 25) calculated a similar, average spacing of 26 feet for the top 6 shorelines in the Bitterroot Valley.

Figure 5.11 Lake Missoula shorelines north of the Missoula Valley etched in relatively soft sediments.

The even spacing, uniformity, and faintness of the shorelines actually make no sense within the multiple flood model in which each shoreline represents a distinct filling of glacial Lake Missoula. If there were 40 to 100 fillings of the lake during the "last" ice age, not to mention the lakes that must have formed in many previous ice ages, there would be a random set of shorelines with large wave-cut notches and abundant material in the shorelines. Each filling of the lake would have mostly eroded the lower shorelines from previous lakes. The only way to account for such uniform and evenly spaced shorelines within the multiple flood hypothesis is if each lake "always" filled to within about 30 feet of the previous lake. The odds of this happening about 35 to 40 times is minuscule. Plus, there should be more debris along each lower shoreline because the lower shorelines would have had more time to erode the rock, but the lower shorelines are no more developed and contain no more debris than the higher shorelines.

It is actually quite possible that the Ninemile rhythmites and the shorelines *are* correlated, but in a different sense than postulated in the multiple flood hypothesis. The Ninemile section would actually represent about 35 to 40 *years* in which the lake continued to fill in relatively deep water above the section. Silt/clay rhythmites require fairly deep water. During this time the lake would be rising up the hills surrounding Missoula. Therefore each shoreline would represent one year of time, mainly a *seasonal stillstand* probably in fall and winter when ice covered the lake. As the lake rose each spring and summer, the rising water would have *protected* the newly formed shorelines underwater. Bretz, Smith, and Neff (1956, p. 1035) also had a similar idea with each shoreline representing a short stillstand that was subsequently covered by rising water:

> The multiplicity and very immature development of Lake Missoula's shore lines argue against a long stillstand at any one level. Although the highest of the series is no stronger than any others, the universal faintness may be due in part to submergence as the basin became filled.

Further evidence for the interpretation of seasonal stillstands is that the shorelines are gently etched in a variety of rock types; there does not appear to be any difference in the shape of the shorelines whether cut in hard Belt Supergroup rocks or in soft clay (Weber, 1972). If each shoreline represented a long stillstand, the soft lithologies would have been more developed with larger shorelines. David Alt (2001, p. 68), in referring to the Missoula Valley, states: "But the shorelines seem just as numerous on the weak clay slopes north of the Missoula Valley as on extremely stable slopes of Belt rocks (Alt, 2001, p. 68). Figure 5.11 shows shorelines etched on the hills of softer rock northwest of Missoula. They appear very similar to those cut on Belt rocks on Mounts Jumbo and Sentinel.

Chapter 6

Giant Ice Age Floods Galore

John Shaw and colleagues (1999, 2000) have summarized evidence that there was only one huge flood from glacial Lake Missoula. But they also claimed this flood was aided by another flood, fifty times the size of the Lake Missoula flood that surged down the Okanogan River Valley underneath the Cordilleran Ice Sheet from interior British Columbia. How can they postulate a flood much larger than the Lake Missoula flood? Is there any evidence for such a flood?

ICE AGE FLOODS DETECTED WHEN UNIFORMITARIANISM IGNORED

Many giant ice age floods are now postulated; a flood from British Columbia is just one of them. I will start with a chronological progression of how scientists have reinterpreted deposits when they temporarily overlooked their uniformitarian bias. Once the scientific community in the 1960s and 1970s agreed that the Lake Missoula flood actually took place, they "opened their eyes" and started to recognize that other ice age floods also occurred in different regions. Deposits thought to be the results of normal glacial processes were reinterpreted as features formed by giant ice age floods. Although scientists had previously examined these particular deposits and saw only slow glacial processes, they looked at them with renewed vision in terms of ice age floods. They discovered that they had overlooked evidence for massive ice age flooding. This was mainly because glacial geologists have been trained to think in terms of present, slow processes–the uniformitarian conceptual box. It is because of this mindset that they missed the Lake Missoula flood and why it

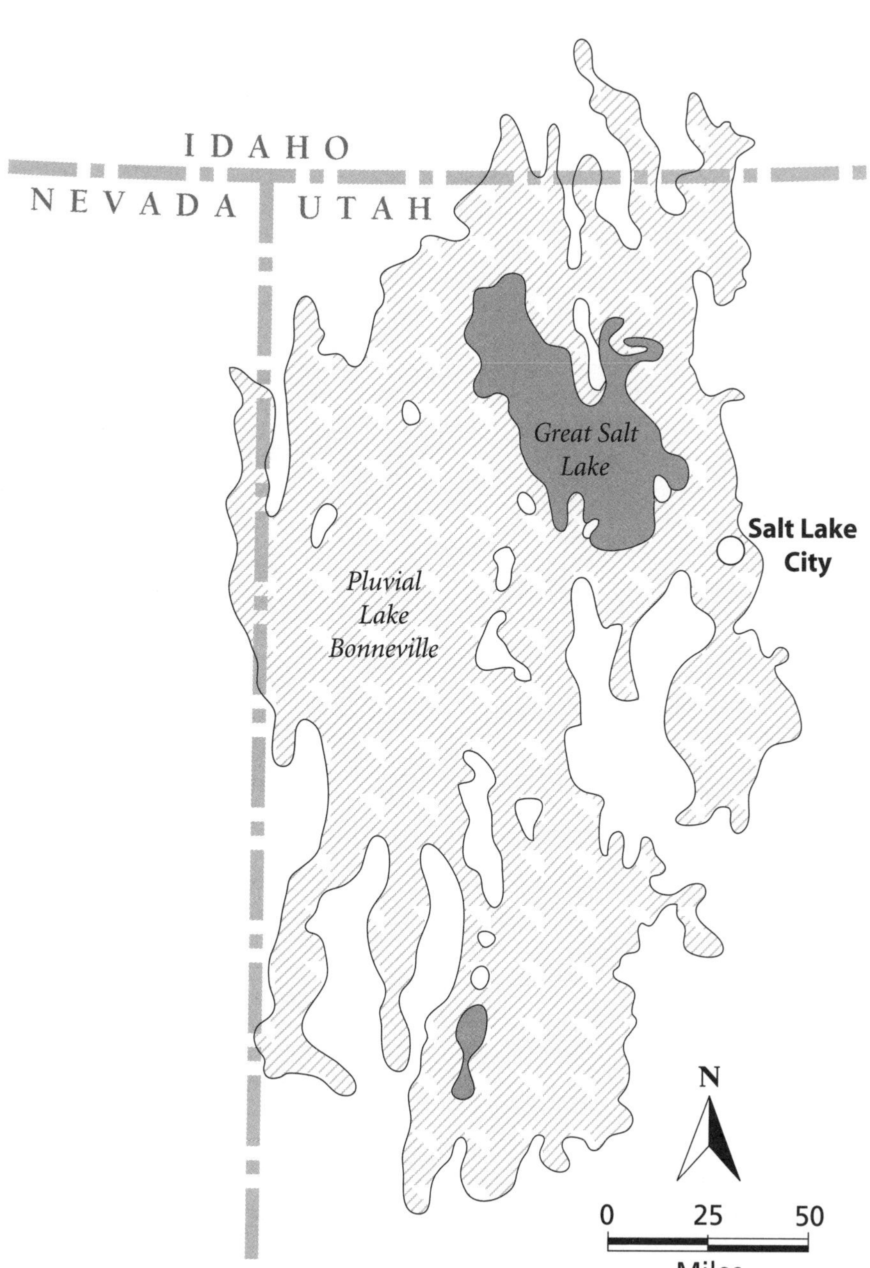

Figure 6.1 Map of pluvial Lake Bonneville (drawn by Mark Wolfe).

Figure 6.2 Shorelines of Lake Bonneville along the Wasatch Mountains, north of Salt Lake City, Utah.

Figure 6.3 Red Rock Pass, Idaho, from where the Bonneville flood initiated.

took 40 years for Bretz to convince scientists of its existence, despite obvious observational evidence.

The size of these other recognized floods started off much smaller than the Lake Missoula flood, but as time went on, newer and larger floods were postulated. The size of the floods seems to have grown with time. This is why Shaw and colleagues (1999, 2000) can now postulate such a huge flood from British Columbia and get the evidence published, although with difficulty. This monstrous flood from British Columbia is just an extension of two other massive floods that Shaw (1996, 2002) and many other researchers during the past two decades have postulated. These two floods, many times the size of the Lake Missoula flood, spilled from underneath the southern edge of the Laurentide Ice Sheet in northeast and north central North America. Is there any evidence for such monstrous subglacial floods? I will delve into this question in the last two sections, but before, I will document the discovery of the recognition of other huge ice age floods.

THE BONNEVILLE FLOOD

One of the first floods recognized after the Lake Missoula flood was accepted is the Bonneville flood that swept down the Snake River in southern Idaho. This flood originated from Lake Bonneville (Figure 6.1), one of the pluvial lakes in the southwest United States that existed during the wet period of the ice age. A small remnant of Lake Bonneville is the Great Salt Lake in Utah. Shorelines from Lake Bonneville are very evident on the hills and mountains around Salt Lake, especially along the Wasatch Mountains (Figure 6.2). We know that the lake and flood were associated with the ice age because its shorelines have been etched on a terminal moraine that was left by an ice stream that emptied from Little Cottonwood Canyon Creek in the Wasatch Mountains. The flood burst when Lake Bonneville rose to its highest point and eroded through an alluvial fan just north of Red Rock Pass in southeast Idaho (Figure 6.3) (O'Connor, 1993). The lake water is believed to have drained over a period of at least eight weeks (Jarrett and Malde, 1987) falling 354 feet (108 meters) to the position of the stable Provo shoreline of ancient Lake Bonneville. It discharged 1150 mi^3 (4750 km^3) of water from Lake Bonneville, three times greater than the volume previously inferred (Jarrett and Malde, 1987). This volume is twice the volume of water released from glacial Lake Missoula, but the discharge was much smaller in the Bonneville flood.

The torrent spread from Red Rock Pass down the Portneuf River and into the Snake River. It rushed down the Snake River of southern Idaho at variable speeds, depending upon the width of the valley and the ability of the water to fan out onto the surrounding Snake River Plain. The floodwaters finally flowed into the Columbia River of eastern Washington, from where it made its way to the sea. The peak discharge, depth, and velocity along the path were calculated by finding the maximum height of erosional and deposition features. It was calculated that in narrow constrictions at maximum flow the discharge was about 35 million ft^3/sec (1 million m^3/sec), the depth up to 400 feet (120 meters), and the velocity up to 92 mph (41 m/sec) (O'Connor, 1993). The flood was much shallower and slower through the wider areas of the Snake River Valley. Its speed and height were comparable to the Lake Missoula flood only through narrow constrictions, but the discharge through these narrows was 1/15 that of glacial Lake Missoula (O'Connor and Baker, 1992).

The Bonneville flood formed a complex of abandoned high-level channels and cataracts. Many bars are of comparable magnitude, structure, and composition to those of the Channeled Scabland of eastern Washington (Bretz, 1969, pp. 530-532; O'Connor, 1993). The flood was fairly shallow as it moved into the Pasco Basin from the Snake River. Since the Bonneville flood discharged over eastern Washington before the Lake Missoula flood, there is no trace left in eastern Washington. The Lake Missoula flood reworked the deposits from the Bonneville flood as expected. The farthest downstream the Bonneville flood deposits are preserved is around Lewiston, Idaho, where Lake Missoula flood deposits overlie those from the Bonneville flood (Baker, 1983, p.123; O'Connor, 1993). It is ironic that Flint's outrageous counter hypothesis to the Lake Missoula flood at one time used the channeled scablands of southern Idaho as evidence that normal water flows could produce the Channeled Scabland of eastern Washington, but these scablands in Idaho actually were the results of the catastrophic Bonneville flood (O'Connor, 1993, p. 63).

COMMONPLACE ICE AGE FLOODS

As the Laurentide Ice Sheet retreated in North America, great amounts of meltwater caused numerous glacial lakes to form as ice blocked their exit. Some of these ponded glacial lakes released water catastrophically, but the details are complex (Teller, 1995). Victor Baker (2002, p. 2379) notes:

> Over the past 40 years, evidence has accumulated for catastrophic failures of ice-dammed lakes, overflows of lakes that had formed along ice margins, and subglacial outburst flooding in the many river systems associated with the immense continental ice sheets of the last ice age.

Large, deeply incised spillways were created during these

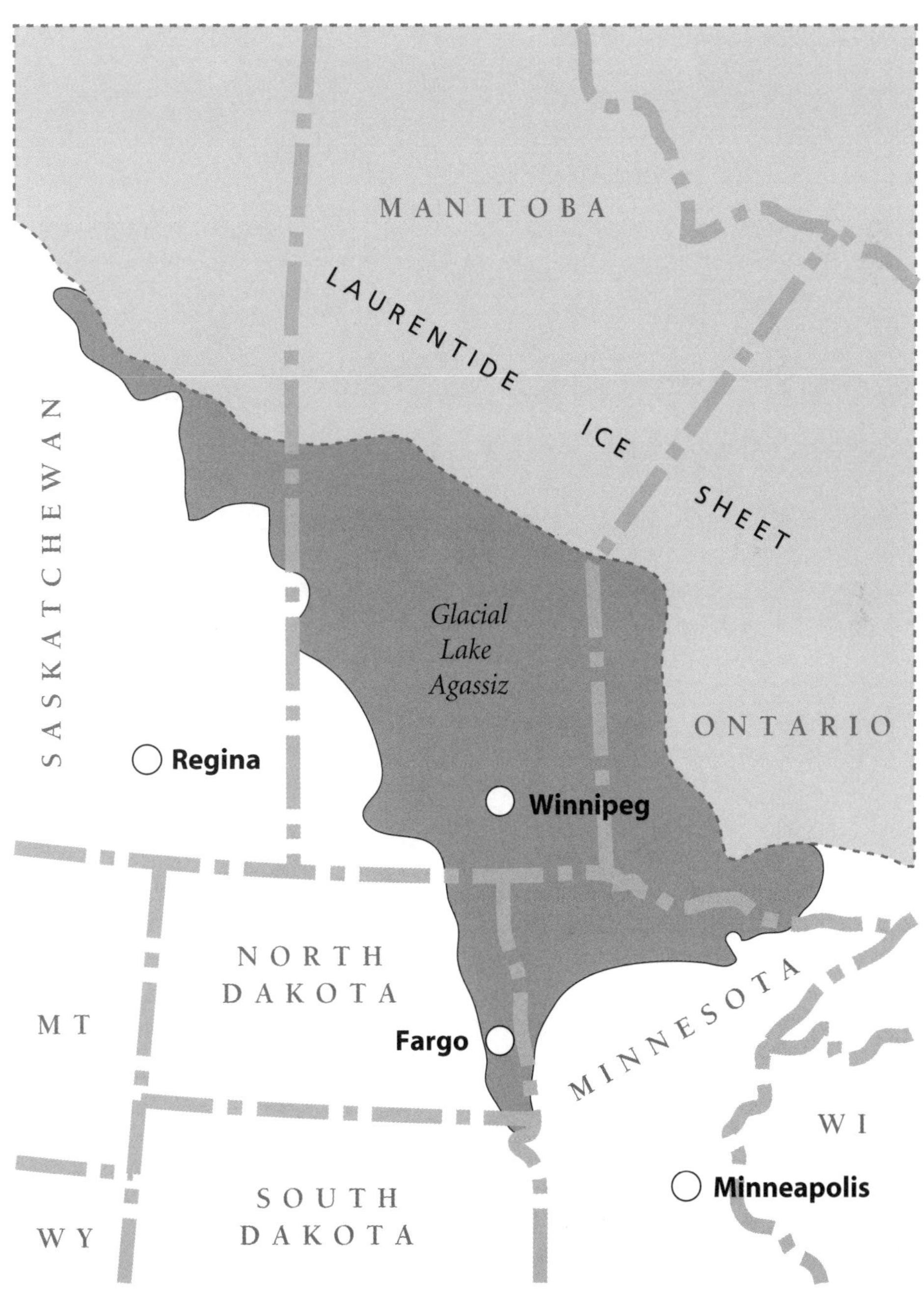

Figure 6.4 Glacial Lake Agassiz of central Canada (drawn by Mark Wolfe).

Figure 6.5 Drumlin from east of Banff, Alberta, Canada

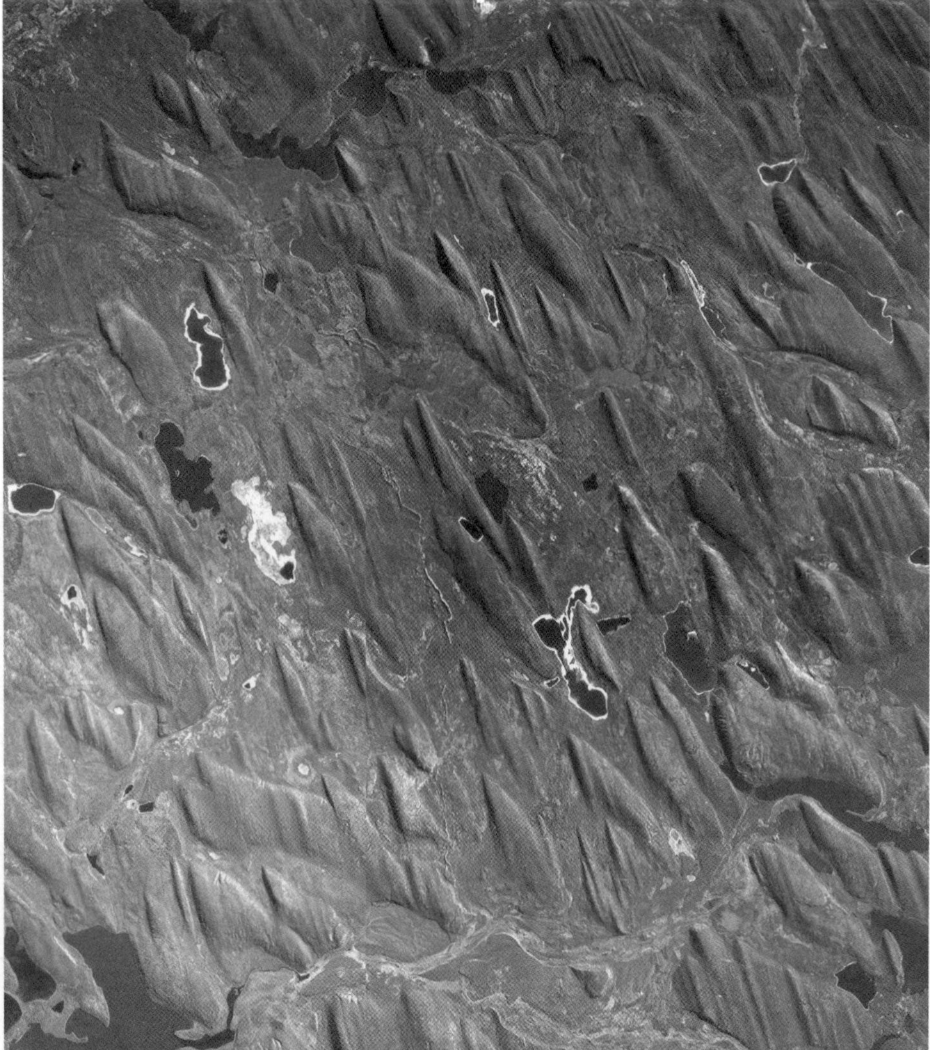

Figure 6.6 Livingstone Lake drumlin field from the air in northern Saskatchewan (courtesy of the Geological Survey of Canada).

events (Kehew, 1982; Kehew and Lord, 1986, 1987; Kehew and Teller, 1994).

The largest ice age lake was glacial Lake Agassiz (Figure 6.4). Lake Winnipeg is a modern-day shriveled remnant of this huge lake. The initial outflow from glacial Lake Agassiz discharged southward through eastern North Dakota and into the Mississippi River and Gulf of Mexico. This great discharge is thought to have formed the large Charleston alluvial fan in southeast Missouri near the junction of the Mississippi and Ohio Rivers (Porter and Guccione, 1994). Further downstream in the lower Mississippi River's alluvial plain, catastrophic flooding caused the river to shift widely and form distinctive relict channels (Teller, 1995). It is now postulated that glacial Lake Agassiz also periodically released catastrophic discharges to the southeast into the Great Lakes (Teller, 1987; Teller and Thorliefson, 1987). The largest of these floods is supposed to have released 1000 mi^3 (4100 km^3) of water into Lake Superior within only a year or two with a peak discharge of 7.1 million ft^3/sec (200,000 m^3/sec). This is much less than that of the glacial Lake Missoula flood, but significant enough to cut channels in bedrock with steep-walled canyons over 300 feet (90 meters) deep. Deposits from these floods may have been detected in southern Lake Michigan (Colman, Keigwin, and Forester, 1994). Glacial Lake Agassiz is also considered to have caused a catastrophic flood northwest into the Mackenzie River Valley of the Yukon Territory with a peak discharge of 85 million ft^3/sec (2.4 million m^3/sec) for 78 days (Smith and Fisher, 1993).

Catastrophic flooding of glacial lakes is reported outside of North America, as well. The mountains forming the border regions of modern Russia, China, Mongolia, and the central Asian republics show abundant evidence of ice-dammed lakes, many of which emptied catastrophically (Baker, 2002; Rudoy, 2002). A large lake was formed in the northwest Altai Mountains, central Asia, by a mountain ice lobe that blocked the Chuja River (Baker, Benito, and Rudoy, 1993; Carling, 1996; Rudoy, 2002). When the ice dam burst, the flood was at least 1300 feet (425 meters) deep with a peak discharge of 636 million ft^3/sec (18 million m^3/sec) and a maximum velocity of 100 mph (45 m/sec). This estimate slightly exceeds the peak flow and velocity of glacial Lake Missoula. Spectacular superflooding also occurred in the upper Yenesei River of western Siberia (Baker, 2002, p. 2379).

Figure 6.7 An esker in Bow Valley, west of Calgary, Alberta, Canada.

Smaller ice age flooding is suggested for portions of the Scandinavian Ice Sheet in northern Europe and northwest Asia. Based on boulders up to 16 feet (5 meters) wide in a large boulder delta complex in Swedish Lapland, a catastrophic drainage of an ice-dammed lake is suggested (Elfström, 1983, 1987). A small ice dammed lake of only about one cubic mile (4 km3) in Scotland is thought to have drained catastrophically. The lake is thought to have formed the distinctive shorelines called the parallel "roads" of Glen Roy (Sissons, 1979).

There are undoubtedly other catastrophic floods associated with the ice age. The major superflooding events from ice-dammed lakes should have been mostly discovered by now. Further small- to medium-sized floods will probably be discovered in the future. All of the above events are generally well documented, but one class of flooding is still quite speculative, and if true would make the Lake Missoula flood seem like a small stream flood. These are the subglacial floods that have been postulated in the 1980s and 1990s by John Shaw and colleagues.

TITANIC SUBGLACIAL FLOODS

The hypothesis of titanic subglacial floods was first suggested by John Shaw in an attempt to explain the origin of atypical drumlins in northern Saskatchewan, Canada (Shaw, 1983, 2002; Shaw and Kvill, 1984). A drumlin is a low, smoothly rounded, elongated hill (Bates and Jackson, 1984, p. 152) (Figure 6.5). The ones in question are the famous Livingstone Lake drumlins on the Athabasca Plain that show spectacular linear forms from the air (Figure 6.6). They are lens shaped and up to about 200 feet (61 meters) high, a mile (1.6 kilometers) long, and about 700 feet (215 meters) wide and are mostly composed of glacial debris that is sometimes stratified. The long axis is oriented in the direction of ice or water flow, as if streamlined. The origin of drumlins is still unknown, since they do not seem to be forming under ice today, as far as anyone knows. Drumlins are another one of those ice age mysteries, and there has been much ink spilled in attempting to explain them. In fact, the origin of ice ages is still an unsolved mystery despite 160 years of work by mainstream scientists (see Chapter 8).

John Shaw noticed that the Livingstone Lake drumlins were similar in form to flutes made on turbidite beds in sedimentary rocks, although over a hundred times larger (Shaw, 1996, p. 185). Turbidite beds are formed by the deposition of turbidity currents. These are currents that flow down a slope under water. A turbidity current deposits a fining-upward sequence including laminations. Shaw suggested the bold hypothesis that the drumlins were formed by meltwater eroding cavities under the ice, carved in much the same way as a river in a cave carves away its roof leaving erosional marks (Shaw, 1983; Shaw and Kvill, 1984). These lens-shaped erosional marks at the bottom of the ice sheet were subsequently filled by glacial debris from flowing meltwater. The debris within the drumlin can be stratified and sorted, indicating deposition by water. Such drumlin genesis implies large volumes of turbulent meltwater moving as a sheet under the ice.

Shaw found further support for his theory of subglacial flooding from other glacial features in the Livingstone Lake area, such as tunnel valleys and eskers that have been formed by water under great pressure. A tunnel valley, also a mysterious phenomenon of the ice age, is a large, elongated overdeepened depression with steep walls cut into bedrock or glacial debris (O' Cofaigh, 1996). They are fairly numerous in

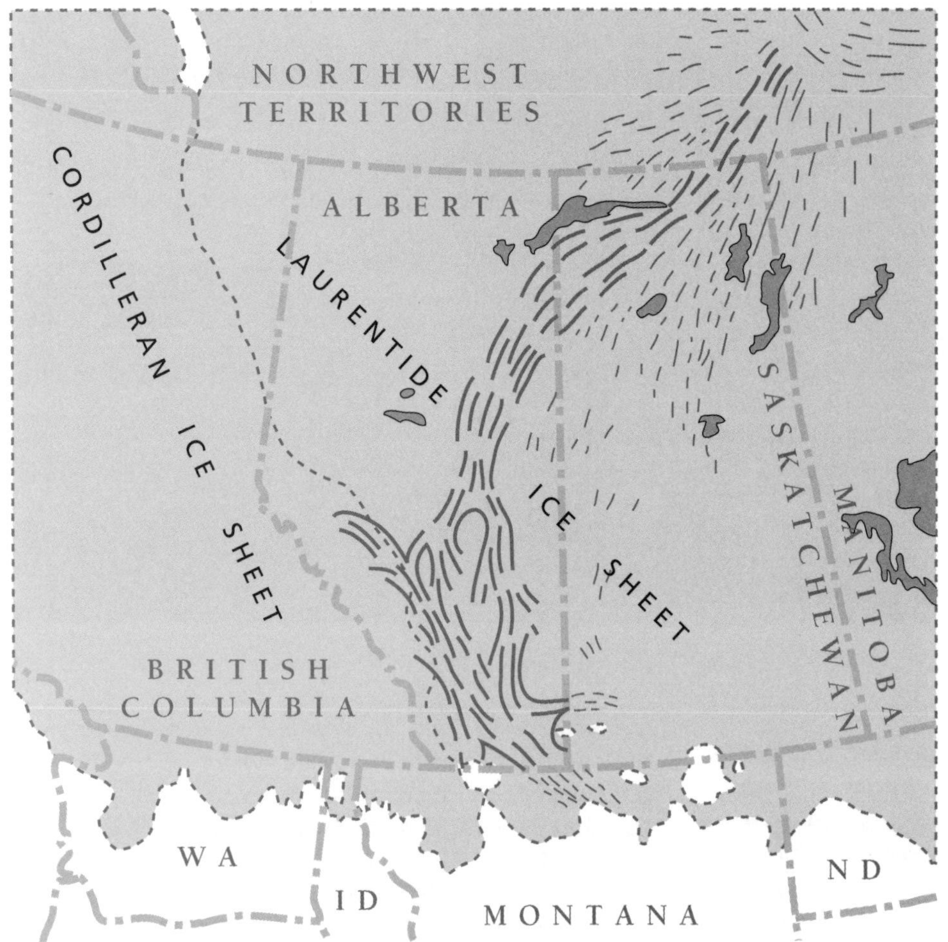

Figure 6.8 Western subglacial flood path that extends from the Northwest Territories, northwest of Hudson Bay, southwest through northern Saskatchewan, and turns south through central Alberta and then southeast through southern Alberta and into northern Montana (redrawn from Shaw, 1996, p. 219 by Mark Wolfe).

glaciated areas and have dimensions on the order of several hundred feet (100 meters) deep, over 60 miles (100 kilometers) long, and up to 2 miles (3 kilometers) wide. The enigmatic tunnel valleys fit nicely into Shaw's hypothesis, since they would be expected to form from channelized flows after transformation from subglacial sheet flows (Gilbert, 1990; Shoemaker, 1992, pp. 1254, 1256; Shaw *et al.*, 1996, p. 1156). Eskers are linear, winding ridges of coarse stratified gravel and sand left behind by a stream flowing within or beneath stagnant ice (Bates and Jackson, 1984, p. 170) (Figure 6.7). They can be more than 300 miles (500 kilometers) long, narrow, and average 100 feet (30 meters) high. Eskers are observed to run up and over low hills and traverse drumlins, indicating that the meltwater traveled under high pressure, and that the eskers were formed after drumlins.

The idea that meltwater can form drumlins has caught on and a fair number of scientists now advocate this hypothesis (Dardis, McCabe, and Mitchell, 1984; Shaw and Sharp, 1987; McCabe and Dardis, 1989; Shaw, 1989; Shaw, Kvill, and Rains, 1989; Shoemaker, 1995, 1999; Bluemle, Lord, and Hunke, 1993; McCabe, Knight, and McCarron, 1999). However, most glacial geologists disagree, probably because of the implications of such flooding. The majority favor plastic deformation of loose glacial debris into ridges during glacial movement, pointing out that many drumlins are composed of unsorted glacial debris. Shaw (1996, pp. 208-218) recognizes that drumlins can be composed of unsorted glacial debris, but many of them are also made up of stratified, water laid coarse gravel, and as a result can have a different origin. The classic drumlins, unlike the Livingstone Lake drumlins, are thought to have been formed by glacial ice erosion or by meltwater flowing under ice. Shaw and colleagues favor the latter hypothesis, especially since the classic drumlins have narrow crescentic troughs that wrap around the upstream end and extend parallel to the long-axis of the drumlin. These troughs are furrows along the flanks of the drumlin. Water seems to be the only plausible explanation for the formation of these crescentic troughs.

Once the origin of drumlins by huge catastrophic subglacial floods was postulated, many researchers suggested that other enigmatic glacial features could be formed by such a mechanism (Shaw, 1996, p. 208). These include flutes, Rogen moraines, and interlobate stratified moraines. A flute is a scoop-shaped depression or groove on bedrock or glacial till. Rogen moraine is composed of small ridges and troughs of debris transverse to glacial or flood water motion (Fisher and Shaw, 1992). Interlobate stratified moraines are moraines between glacial lobes that contain stratified glacial debris. Shaw and colleagues present an impressive array of evidence for subglacial flooding. Water scour marks, such as sichelwannen (small sickle-shaped erosional depressions with sharp upstream rims) are sometimes found in glaciated terrain and are considered to be caused by water erosion (Shaw, 1996, pp. 186-200). They are especially numerous along the northeastern shore of Georgian Bay of Lake Huron, Ontario. These features formed under the ice because some of them are lightly striated. Included in this group of micro landforms are potholes, sometimes cut on the sides of anticlines (Gilbert, 2000). The potholes sometimes show percussion marks on their walls, which were likely caused by high-pressure bubble collapse during cavitation by high speed, shallow currents, just as one would expect underneath an ice sheet.

Percussion marks are semi circular marks on the rock (see Figure 7.17). Most impressive are large-scale flutes in bedrock, as seen from the air downflow from rock escarpments (Shaw, 1988a, 1996; Tinkler and Stenson, 1992). Their regularity and other features indicate water origin. Catastrophic subglacial flows can explain a wide variety of previously enigmatic glacial bedforms:

> The meltwater hypothesis for underice bedforms is not rejected by similar inductive tests. Rather, the hypothesis is found to explain an increasingly wide variety of otherwise enigmatic bedforms and to combine them successfully in a single model (Shaw, 1996, p. 186).

Two main subglacial flood paths have been deduced from all the evidence, which could be a simplification. One starts in the Northwest Territories, northwest of Hudson Bay, traverses southwest through northern Saskatchewan, forming the Livingstone Lake drumlin field, and then turns south through central Alberta and then southeast through southern Alberta and into northern Montana (Rains *et al.*, 1993, 2002; Sjogren and Rains, 1995; Munro-Stasiuk, 2000; Shaw *et al.*, 1996, 2000; Munro-Stasiuk and Shaw, 2002). This flood path is shown in Figure 6.8. A minor flood path from this major path has been postulated from the Northwest Territories westward through the southern Slave Province (Rampton, 2000), but this suggested path has not been as well studied yet. The many unique bedforms are indicative of catastrophic flow. Northwest-southeast channels cut in the continental divide between the Del Bonita Uplands and the Cypress Hills of southern Alberta are considered subglacial tunnel valleys formed during catastrophic flow (Beaney, 2002). Transverse bedforms southeast of the tunnel valleys are reminiscent of ripple marks (Beaney and Shaw, 2000). One powerful evidence for this flow is the ubiquitous presence of large granite, red-granite migmatite, gneiss, and other crystalline erratic boulders in southern Alberta and northern Montana (Figure 6.9). The nearest source of these erratics is in northern Alberta or central and northern Saskatchewan, which is part of the Canadian Shield. These erratics are more than 10 feet (3 meters) in length and most are rounded to subrounded, reflecting water action, a characteristic also noted by Beaney and Shaw (2000, p. 54). Some rocks have percussion marks that point to high current speeds: "Percussion marks are indicative of clast-to-clast contacts in a highly energetic fluvial environment" (Beaney and Shaw, 2000, p. 57). It is of interest that the many landforms thought to have been molded by a subglacial flood are unchanged by further ice age processes (Shaw *et al.*, 2000, p. 978) indicating that after the flood, the ice was either stagnant or the subglacial flood broke up the ice sheet along its flow path in Alberta and northern Montana. Shaw and colleagues have not considered this latter possibility.

A second major pathway for a giant subglacial flood is postulated to extend from either Hudson Bay or central Labrador south through the eastern Great Lakes and into the United States (Sharpe and Shaw, 1989; Shaw and Gilbert, 1990; Shoemaker, 1992; Gilbert and Shaw, 1994; Brennand, Shaw, and Sharpe, 1996; Kor and Cowell, 1998). The flow path appears to diverge in southern Ontario and New York. It is claimed that this subglacial flood carved tunnel valleys, produced drumlin fields, and excavated the New York Finger Lakes (Gilbert and Shaw, 1994: Shaw, 1996, p. 222). Kor and Cowell (1998) show many remarkable features on aerial photographs from the Bruce Peninsula, Ontario, indicating a southwest flowing sheet current. Large rounded rocks of exotic origin up to 6 feet (2 meters) in diameter are common on the Bruce Peninsula (personal observation). Some of these rocks also display percussion marks (Kor and Cowell, 1998). This is impressive evidence for subglacial flooding.

What is the magnitude of the catastrophic flood discharge in the two pathways and where did the water originate? These are perplexing questions, but ballpark figures for the rate of discharge can be calculated by the width of the water-eroded bedrock, the height of drumlins, and other features. In the Livingstone Lake drumlin field, Shaw (1996, 2002) conservatively estimated its width at about 94 miles (150 kilometers), its depth about 130 feet (40 meters), and a flow velocity of

Figure 6.9 Ten-foot-long gneiss erratic from Turner, north central Montana, a few miles south of the Canadian border. This boulder likely was deposited from a subglacial flood. This erratic is subrounded to subangular; most boulders are rounded to subrounded.

22 mph (10 m/sec). Based on this information, Shaw (1996, pp. 218-227) estimates that the discharge along the western flood path is on the order of 2.1 billion ft^3/sec (60 million m^3/sec), approximately *four* times the peak discharge of the Lake Missoula flood. Based on assumptions of the volume of ice melted to form the Livingstone Lake drumlin field, Shaw estimated the total volume of water discharged as 20,500 mi^3 (84,000 km^3). This volume represents about *40 times* the volume of glacial Lake Missoula. Moreover, this flood needed enough pressure to force the water uphill for long distances. Shaw tentatively assumes these same values for the eastern flood path. Interestingly, he considers all these values as *underestimates*. The implications of subglacial floods south of the ice sheet, if they occurred, are only beginning to be explored.

Since the subglacial floods originated far to the north, the water source lies in the vicinity of Hudson Bay (Shaw, 1996). Shaw postulates that a lake developed on top of the ice from pooling of meltwater (Figure 6.10), similar to what we see forming on small, flat areas of some glaciers today (Shaw, 1996, p. 225). The water gradually percolated down to the base of the ice sheet. Once at the base, the lake water can flow under the ice becoming highly pressurized by the height of the water in the lake. This is much the same way as an elevated water tank provides water pressure for our water faucets.

There is one major problem with Shaw and colleagues' hypothesis. The huge volume of meltwater required for the subglacial floods originated from near the supposed *center* of the ice sheet around Hudson Bay. This is what many scientists find objectionable (Piotrowski, 1987; Boyce and Eyles, 1991, 2000; Hindmarsh, 1998). During deglaciation, the center of the Laurentide Ice Sheet is thought to be too thick and cold to form even a small fraction of the meltwater Shaw proposes. Melting, they think, was confined to the edge of the ice sheets. Shaw's scenario is radically different from the standard glacial theory of one huge ice sheet centered on Hudson Bay, or the revised glacial model of several domes and a thinner ice sheet. Shaw and colleagues' hypothesis turns the current mainstream ideas of the ice age on its head (see Chapter 8 for an alternative view of the ice age). This likely is the main reason that most glacial geologists reject what is considered to be another *outrageous hypotheses*. In fact, Shaw and colleagues are in a similar position as J Harland Bretz after he finished most of his fieldwork on the Lake Missoula flood in eastern Washington.

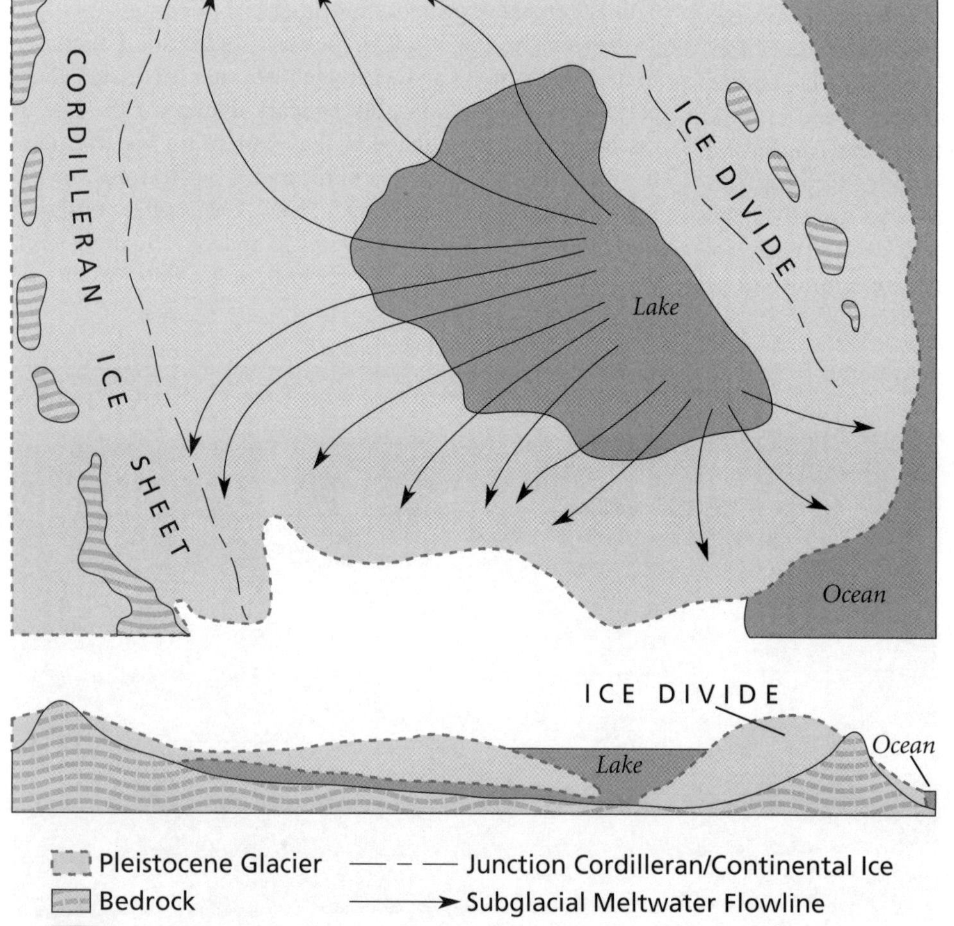

Figure 6.10 Postulated location of the lake in the vicinity of Hudson Bay and subglacial flow paths from this lake. A cross section along the line shown in the top diagram is presented in the bottom diagram (redrawn from Shaw, 1996, p. 226 by Mark Wolfe).

A MONSTROUS FLOOD FROM BRITISH COLUMBIA?

It was within the titanic subglacial flood paradigm that Shaw and colleagues extended their research to a subglacial flood underneath the Cordilleran Ice Sheet. It is this flood that combines at the same time with the Lake Missoula flood to produce the unique features of the western Channeled Scabland. The main reason for their belief in such a monstrous flood from British Columbia is they do not think the Grand Coulee could have formed by only one Lake Missoula flood (Shaw *et al.*, 2000, p. 607). Bretz (1969, p. 527) felt the same way. Whether this is true has not been demonstrated. The flood from British Columbia is commonly thought to have contained *50 times* the amount of water that was discharged from glacial Lake Missoula. In a letter-to-the-editor reply to the study of Shaw and colleagues (1999), Komatsu and others (2000) add that a computer model of the Lake Missoula flood did not have nearly enough water to flood the western scablands to the height deduced from high water marks. However, computer models are simplistic interpretations of reality, and as a result may be in error. It is difficult to evaluate Shaw and colleagues' hypothesis. There is some evidence in the form of possible meltwater-formed drumlins in central British Columbia (Shaw *et al.*, 1999, p. 607). Okanagan Lake in southeastern British Columbia has been glacially overdeepened and is essentially an inland fjord. (Note that Okanagan is the Canadian spelling and Okanogan the United States spelling.) There are up to 2600 feet (800 meters) of ice age sediments in this long north-south lake (Eyles, Mullins, and Hine, 1990, 1991). The relief between the bedrock below the sediments and the surrounding plateau is 6500 feet (2000 meters). This difference is greater than at Grand Canyon. All the sediments in this deep trough accumulated during the *deglaciation of the last ice age* in the uniformitarian multiple ice age scheme. The deep trough is thought to have been dug by an enormous amount of pressurized meltwater (Vanderburgh and Roberts, 1996). The lower layers in the trough are composed of boulder gravel, consistent with such a glaciofluvial mechanism. The fact that the trough is V-shaped and not U-shaped adds additional support for the meltwater origin rather than by glacial ice. Shaw and colleagues (1999, p. 607) simply point out that such high magnitude drainage must have passed south into Washington and extended into the western Channeled Scabland. So, a flood originating from British Columbia is plausible, but other researchers of the Lake Missoula flood have not yet adjusted their thinking to such a possibility.

Chapter 7

The Largest Flood of All– The Genesis Flood

The Lake Missoula flood controversy is a lesson on how certain scientific biases about the past intrude when new ideas or data are introduced. The public expects scientists to be objective, carefully weighing all the possibilities before coming to conclusions. We do not expect a scientist to censor ideas with which he disagrees. But instead he is to use hard evidence to prove or disprove them. David Alt (2001, p. vii) ponders what the dispute over Bretz' hypothesis means in regard to the condition of science in general:

> This is also a story of scientists grappling with an emerging science controversy. Some handled it well, others miserably as personalities, pride, and outright prejudice superceded scientific evidence. This is not how science should work, but how it often does work.

The public perception of science is mostly a myth.

Despite the obvious signs of a gigantic flood, the leading lights in geology opposed Bretz and his hypothesis with *vehemence* (Baker, 1983, p. 122). Most of them went to their graves still believing that he had gone too far. They feared he had committed the ultimate scientific sin by straying from "good science" into the realm of biblical catastrophism, thereby rejecting 100 years of "enlightenment." In their mind's eye, a flood of biblical proportions was reminiscent of the so-called dark ages when most scholars believed Noah's Flood produced the earth's rocks and fossils. E. C. Olson (1969, p. 503) pontificates: "Geology was not a science until the legendary Noachian flood and six-day creation were replaced by explanations derived from careful study of the rocks." The "careful study of the rocks," however, has been accompanied by a straightjacket mentality of how they should be interpreted. This explains why controversy raged over the Lake Missoula flood–it challenged the dogma of uniformitarianism. Bretz (1978, p. 1), himself, acknowledges this:

> Catastrophism had virtually vanished from geologic thinking when Hutton's concept of "The Present is Key to the Past" was accepted and Uniformitarianism was born. Was not this debacle [the Spokane flood] that had been deduced from the Channeled Scabland simply a return, a retreat to catastrophism, to the dark ages of geology? It could not, it must not be tolerated.

The 40-year history of denial shows how geologists were willing to ignore the obvious, often refusing to even *examine the evidence in the field* before denying the reality of the Lake Missoula flood. They even invented other "outrageous hypotheses" to counter Bretz's claim.

Baker (1978a, p. 15) shows how uniformitarianism and an anti-Flood bias clouded the minds of the Lake Missoula flood skeptics:

> One cannot but be amazed at the spectacle of otherwise objective scientists twisting hypotheses to give a uniformitarian explanation to the Channeled Scabland. Undoubtedly these men thought they were upholding the very framework of geology as it had been established in the writings of Hutton, Lyell, and Agassiz.

It goes without saying that most geologists are biased against the largest postulated flood of all–the universal Genesis Flood as described in the Bible. Based on such an overwhelming prejudice, we are left with a tantalizing question: Since the geological establishment rejected the Lake Missoula flood despite copious evidence, is their knee-jerk rejection of the Genesis Flood in spite of abundant evidence for such a Flood? The answer to this question is an emphatic yes, as we shall see. Furthermore, those scientists immersed in the uniformitarian tradition cannot even comprehend the evidence for the Genesis Flood because of their mindset.

But first, we must examine what took place in the early 1800s to see whether the weight of objective geological arguments really caused the rejection of the Genesis Flood in favor of uniformitarianism, as so many people think.

UNIFORMITARIANISM NEVER VANQUISHED CATASTROPHISM

Before uniformitarianism became the accepted *assumption* in earth sciences, intellectuals believed a global Flood laid down most of the sedimentary rocks along with their fossils. This was called catastrophism as the origin of the rocks and fossils. The principle of uniformitarianism took over in the early and mid 1800s through the efforts of James Hutton and Charles Lyell. These men are and have been depicted as paragons of reason that vanquished the forces of superstition, men who based uniformitarianism on copious, objective fieldwork (see quote by Olson above).

Staunch evolutionist, Steven J. Gould, wrote in his book, *Time's Arrow, Time's Cycle*, that the *opposite* is actually true. Charles Lyell was a lawyer with a violent antagonism toward the Genesis Flood. Lyell, as well as Hutton, had a personal vendetta according to Gould (1987, pp. 9-10): I shall try to show that Hutton and Lyell, traditional discoverers of deep time in the British tradition, were motivated as much (or more) by such a vision about time, as by superior knowledge of rocks in the field. Indeed, I shall show that their visions stand prior–logically, psychologically, and in the ontogeny [development] of their thoughts–to their attempts at empirical support.

In other words these men had already decided the earth was old and that the Bible had it all wrong *before* they examined the rocks and fossils. They did not deduce uniformitarianism from the objective consideration of field data; they worked to find field data that appeared to support their *a priori* beliefs.

Gould (1987) even states that Lyell was not even much of a geologist, and that his catastrophist rivals were the truly careful scientists. Martin Rudwick (1990) considers Lyell's extremely influential book, *Principles of Geology*, one long lawyer's brief. (Lyell actually was a lawyer who had turned to geology only seven years before writing his monumental work.) In other words, his book is one long rhetorical treatise. We can conclude that uniformitarianism was not based on factual data or superior reasoning ability but is a subjective assumption or personal choice. Unfortunately, it continues little questioned to the present day for very nonscientific reasons, as Alt (2001, p. 18) states in reference to the Lake Missoula flood controversy:

> Very few scientists in any discipline come to daring new conclusions as they consider the evidence. Most go through life believing what they learned in college, resenting challenges to their settled beliefs, and disliking those who present them. They muddle their way through their problems in a fog of confusion just like people who are not scientists. That may not be how science should be done, but that is how it very often is done.

This candid comment, a "testimony," bears witness to why uniformitarianism continues as geological dogma to this day, and why such strong resistance to catastrophism that exists within individual scientists and their universities–despite the evidence. Uniformitarianism continues to have a stranglehold on the minds of students and professors alike, even despite the recent trend towards local, large catastrophes (neocatastrophism), such as the Lake Missoula flood and the asteroid bombardment hypothesis for the extinction of the dinosaurs.

Since the times of Hutton, Lyell, and Agassiz an immense amount of geological data has been gathered. It is a good time to reopen the question of whether there really was a global Flood.

A BIBLICAL GEOLOGICAL MODEL

A renewed interest in the Genesis Flood as a global event has been developing since about 1960. Many books have been written by scientists defending the Flood as a reasonable theory (Whitcomb and Morris, 1961; Coffin, 1983; Austin, 1994a; Froede, 1998; Roth, 1998; Woodmorappe, 1999a). The renewed interest rose partly because the old uniformitarian paradigm has not been able to explain many aspects of geology. A fresh look at the rocks and fossils by independent thinkers, unencumbered by strict uniformitarianism, is uncovering a remarkable amount of evidence for a global Flood. Today even some uniformitarian thinkers called neocatastrophists are accepting that scattered catastrophes occurred over the course of earth's history. Large meteorite impacts are just one example of events that are now used to account for unexplained features found in rocks and fossils.

In the process of reevaluating the data, a geological model or classification of the Flood was developed by Tas Walker (1994). Since well-written history is the only direct evidence of the past, he used the book of Genesis as the basis for his model, without regard to ideas about geology or the fossils. He and others found that a global flood does an excellent job of explaining many geological features (Walker, 1996a,b; Oard, 2001a,b). Carl Froede (1995) has also developed a geological timeframe based on the Bible, similar to Walker's classification. Since the global flood model is relatively new, there are mountains of data left to be reexamined, but enough has been accomplished to give those who believe in the Flood much encouragement.

Walker's model divides the Flood into two great stages: the Inundatory Stage and the Recessional Stage (Figure 7.1). The Inundatory Stage has three phases and the Recessional Stage two phases. The Inundatory Stage is the time when the water rose and flooded the entire earth (it must be remembered that the pre-Flood earth could not possibly have the same

EVENT/ERA	STAGE	DURATION	PHASE
FLOOD (371 DAYS)	RECESSIVE	100 DAYS	DISPERSIVE
		200 DAYS	ABATIVE
	INNUNDATORY	30 DAYS	ZENITHIC
		20 DAYS	ASCENDING
		10 DAYS	ERUPTIVE

Figure 7.1 Tas Walker's geological time scale based on Genesis only.

Figure 7.2 Grand Canyon from Desert Viewpoint. Note the thick deposits of horizontal strata. The strata tilted eastward at the bottom of the horizontal strata are Precambrian strata thought by some to be pre-Flood strata and by others as Flood strata that has been tilted and sheared off forming an unconformity.

geography and topography we see today because of all the Flood sedimentation and tectonics). The Genesis Flood is described in the book of Genesis as a global flood that lasted one year.

Some have wanted to dismiss it as a local flood in the Mesopotamian Valley or the Black Sea for one reason or another, but the text of Genesis makes it very clear that it was a global Flood. There are many reasons for this. Local floods do not last an entire year; the Bible says the Flood lasted 371 days. If it were a local flood, there would be no need for an Ark to save two of each kind of animal. The animals can simply flee to the mountains. Boats in a local flood would automatically be carried downstream, possibly all the way to the ocean, not end up in the "*mountains* of Ararat." An Ark landing in the mountains would be logical only if it was a global flood. The covenant of the rainbow, God's promise to never flood the earth again, would make no sense if it was just a local flood. There have been countless local floods since then, such as the 1970 hurricane storm surge that killed nearly half a million people in Bangladesh (Oard, 1997d, p. 51). Lastly, the language of the text indicates the whole earth was flooded, and all of the people drowned except Noah and his family. The words *whole* and *all* are very inclusive.

After the Inundatory Stage of the Flood came the Recessional Stage. This represents the retreating of the Floodwaters from the future continents. It was at this time the present geography developed. Walker divides the Recessional Stage into two phases: 1) the abative or sheet flow phase, and 2) the dispersive or channelized phase. These two phases are distinguished by the horizontal scale of the currents. When the Flood first started receding from a globally flooded earth, the currents would have been wide, maybe on the scale of a thousand miles (1600 kilometers) or greater. As more and more land became exposed with time, the water would be forced to flow between newly risen mountain ranges and plateaus, carving channels that would end up as valleys. The second phase of the Recessional Stage continued until the Floodwaters finally drained from the continents.

There is some controversy on the exact details of the Flood as derived from Genesis, but Scripture seems to indicate that the Inundatory Stage lasted 150 days and the Recessional Stage 221 days. This is in line with the 21 weeks of "prevailing" and the 31 weeks of "assuaging" in Whitcomb and Morris's landmark book *The Genesis Flood*. These main events of the Flood are expected to produce certain geological signatures on the surface of the land. In chapter 9, I will show how the Lake Missoula flood can be used to help geologists recognize several geological features that resulted from the Genesis Flood, but which remain very difficult to explain within the uniformitarian paradigm.

VAST SEDIMENTARY LAYERS LAID DOWN DURING THE INUNDATORY STAGE OF THE FLOOD

Walker's biblical geological model is very helpful in understanding the progression of the Flood. A global Flood would cause catastrophic erosion and rapid sedimentation on a vast scale during especially the Inundatory Stage. Sedimentary rocks cover 75% of the Earth's surface and are about 1 mile (1.6 kilometers) deep. They provide us with a crucial test for whether uniformitarianism or catastrophism is the more likely theory. Over the entire earth many layers of rock can be traced over thousands or hundreds of thousands of square miles (many thousand to several hundred thousand square kilometers) (Ager, 1973; Roth, 1998, pp. 218-219). A three dimensional analysis shows these layers are laid down in sheets. The horizontal strata of the top 4000 feet (1220 meters) of Grand Canyon are a good example of sedimentary rocks covering a large area (Figure 7.2). The strata are composed of generally even layers that can be traced for over 200 miles (320 kilometers) in the canyon. They often extend well beyond the canyon (Austin, 1994c). There does not seem to be any present processes, even on the bottom of the ocean that can cause such even layering over such distances.

The Shinarump Conglomerate, considered a terrestrial deposit, is a widespread layer of sand and rounded rocks in the Colorado Plateau that is about 50 feet (15 meters) thick and covers more than 100,000 square miles (256,000 square kilometers) (Brand, 1997, pp. 222-223). The conglomerate is rounded by the action of water. Today, there is no place on dry land that a uniform thickness of sand and gravel is being deposited over a fairly level area anywhere near this size. It seems more likely that the sand and gravel was spread as a thin sheet by at least a regional scale cataclysmic flow of water.

When we carefully examine the vertical dimensions of Grand Canyon's sedimentary rock, keeping in mind their great uniformitarian "age," a contradiction arises. Often, two adjacent, vertical layers of strata are dated millions of years apart, and yet there is little if any sign of erosion between and within layers. According to the evolutionary time scale, the flat lying rocks of the Grand Canyon represent deposition that occurred over a period of 320 million years (the dates are deduced primarily by the fossils found in the rocks). If 4000 feet (1220 meters) of strata were laid down over such a long period of time, where are the signs of erosion between and even within the layers? The lack of erosion in the Grand Canyon deposits contrasts with present erosion rates that would level North America to sea level in about 10 to 50 million years, depending upon whether allowance is made for the activities of man and the decreasing erosion rate as the continents are lowered (Schumm, 1963; Roth, 1998, pp. 263-266). Erosion should have cut a multitude of valleys and canyons through the sedimentary rocks, especially when we consider that the Grand Canyon, itself, which was, as the story goes, eroded in solid rock over a period of only several million years (Lucchitta, 1990). Some may appeal to the marine nature of the rocks in the Grand Canyon, but, even if they were formed on the bottom of the ocean, the ocean bottom is not the quiet place scientists once envisioned. Channels are commonly cut into sheets of sediment that lie on the ocean bottom (Stevenson, Scholl, and Vallier, 1983; Gardner, Field, and Twichell, 1996). So, the idea that these sedimentary rocks were laid down on the bottom of the ocean with no erosion between layers separated by millions of years is doubtful.

The Flood paradigm predicts very little erosion would take place while a large proportion of the sediments were quickly deposited during the Inundatory Stage. The lack of erosion between layers is no surprise, but is actually an expectation of the Genesis Flood. During the Genesis Flood the sediments would be laid down rapidly as sheets over large areas (Roth, 1998, pp. 222-229). As in the Grand Canyon, the rest of the sedimentary rocks across the world show little if any erosion within and between the layers. According to Flood geology, the main erosion of valleys and canyons, as well as faulting and uplift, occurred *after* the entire mass was mostly laid down. In Figure 7.2 the layers on the south side of the Grand Canyon can be matched with those on the north side. All these layers were first laid down horizontally. Then the Grand Canyon and its side canyons were cut afterwards. The evidence from sedimentary rocks strongly favors a global Flood, not slow processes taking millions of years.

THE FLOOD–AN ENVIRONMENT FOR RAPID FOSSILIZATION

Fossils in the sedimentary rocks are very important to the evolutionary time scale. They provide a dating scheme for about 600 million years of supposed time. Not many people realize that trillions of fossils in the rocks point more towards a global Flood than toward evolution and uniformitarianism.

First, fossilization requires rapid burial (Raup and Stanley, 1978, pp. 14-25). The rapid burial of large animals, such as dinosaurs, requires an especially large quantity of sediment. Considering the trillions of fossils, what better mechanism is there than a global Flood for producing them? With rapid burial, the organisms would not have a chance to be eaten or decay on the surface of the earth or the bottom of the ocean. One would think that shelled creatures could lie on the ocean bottom for many years providing more opportunity for burial. Raup and Stanley (1978, p. 15) state that even mollusk shells disintegrate rapidly, even in favorable environments for shelled organisms:

> As soon as an oyster or other mollusk dies, its shell is subject to deterioration resulting from attack by a great variety of boring organisms, including worms, sponges, other mollusks, and algae. Most sea bottoms on which living shelled organisms are abundant have surprisingly few intact, empty shells.

The great majority of fossils are the shells of marine organisms, like clams. There are probably trillions of them in the sedimentary rocks of the earth. These shells also show evidence of rapid burial. In the fossil record, mollusks commonly are found whole–both shells are attached. Since the muscles holding the two shells together decay very quick after death, the shells would have opened in a short time. If burial were by slow processes over vast eons, the shells of organisms like brachiopods and clams should be single at best. But since so many of the shells are attached, burial was rapid.

Even the existence of mollusks with just one shell is evidence of rapid burial. If the organism were somehow protected from scavengers, the shell itself would degrade rapidly in a short time, since the shell is held together by a network of *organic* tissue. Organic tissue disintegrates rapidly after death.

Second, the organism generally must turn to stone in the sediments. This is probably more difficult than burying the organism rapidly. Once buried, organisms generally decay quickly by a variety of biological and chemical decomposition agencies operating in the sediment. And even if protected from these agencies, the organism still must be fossilized. The organic matter has to be replaced or impregnated by inorganic matter in a process called permineralization. (The other fossilization mechanisms, like carbonization, will not be discussed.) The most common mineral substances for permineralization are calcium carbonate and silica (Pinna, 1985, p. 13). (These are, by the way, also the most common cementing agents for sedimentary rocks.)

When we look at modern ground water processes, we find that ground water is low in silica needed for fossilization (Pettijohn, 1975, p. 242). Carbonate is more common than silica in ground water but still much too low to be a significant factor in fossilization. The source of all the carbonate required to cement just a sheet of sand, for instance, is one of many problems associated with carbonate cementation (Pettijohn, 1975, pp. 244-245). All the above factors are probably why permineralization is exceeding rare today. Uniformitarianism is left without a good explanation for rapid fossilization of billions of organisms.

On the other hand, the Genesis Flood would have provided an ideal mechanism for the rapid burial and fossilization of a great number of organisms. As sediments were deposited hundreds of feet at a time, the pressure of the higher sediments on the lower sediments would tend to squeeze the water out of the pores between sediment particles. This water would become highly pressurized the deeper the sediments are buried. The water would also become charged with chemicals as it dissolved the soluble minerals of the sediments. This chemically charged water under high pressure would not only rapidly fossilize buried organisms, but also the sediments themselves.

GREAT VERTICAL UPLIFT AND SUBSIDENCE TO DRAIN THE FLOOD WATER

During the Recessional Stage of the Flood, the Earth had to change from being totally flooded to the present day topography. For the runoff to happen there had to be differential uplift and sinking of the Earth's crust and mantle. Mainstream geologists believe that shifting in the earth's crust is how mountain ranges and adjacent valleys and basins were formed. The Beartooth Mountains are a good example. They are granitic mountains located in south central Montana and north central Wyoming. Granite Peak is the highest peak at 12,799 feet (3902 meters) ASL. In the adjacent Bighorn Basin to the east, granite is located about 10,000 feet (3050 meters) below sea level with sedimentary rocks piled on top of the granite up to about 4000 feet (1220 meters) ASL (Figure 7.3). Along the east face of the Beartooth Mountains, there is close to a 23,000-foot (7000 meters) difference between the granite at the top of the mountains and the granite buried in the valley. This is represented by the Beartooth Fault, indicating 23,000 feet (7000 meters) of nearly vertical displacement. Such differential vertical motion has been demonstrated to be a common tectonic feature of the Earth's crust (King, 1983).

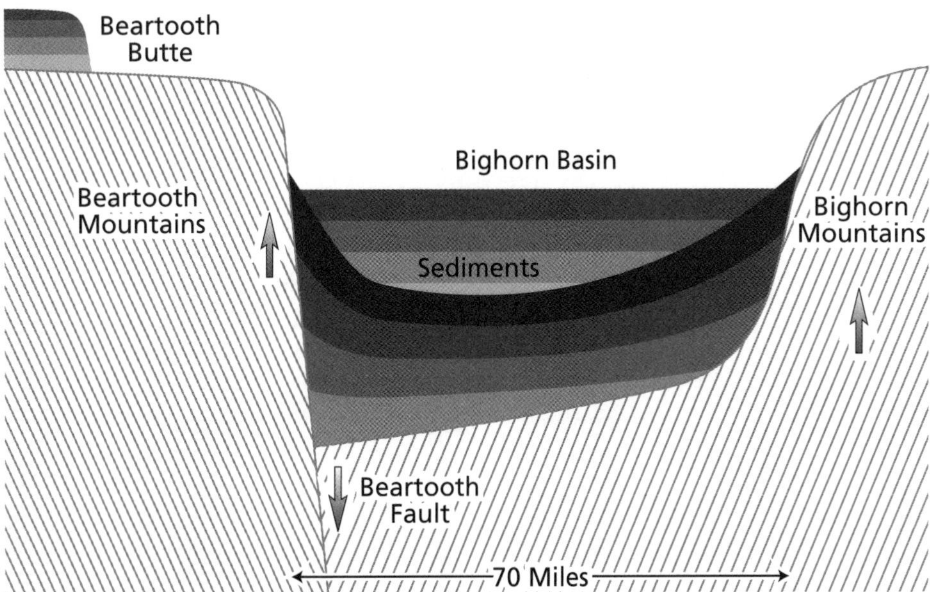

Figure 7.3 Schematic of Beartooth fault showing about 23,000 feet (7000 meters) of vertical displacement (taken from Coffin, 1983). This theme of differential vertical tectonics is very common on the continents and ocean bottom.

This is probably what Psalm 104:6-9 (New American Standard Version) speaks about after the Earth was totally flooded:

> Thou didst cover it [the Earth] with the deep as with a garment;
> The waters were standing above the mountains.
> At Thy rebuke they fled;
> At the sound of Thy thunder they hurried away.
> *The mountains rose*; *the valleys sank down*
> To the place which Thou didst establish for them.
> Thou didst set a boundary that they may not pass over;
> That they may not return to cover the earth [emphasis mine].

Note that the flooding Scripture states is the last time the Earth was globally covered by water. This is clear because the last verse says that God set a boundary that the oceans may not return to cover the Earth. These verses then refer to the Genesis Flood and not to the Third day of Creation when the Earth was first covered by water. Notice too that the water covered the mountains and then hurried away, or drained rapidly, by the *mountains rising and the valleys sinking down*. Some versions of the Bible, such as the NIV, translate this key verse as, "They [the waters] flowed over the mountains, they went down into the valleys." However, this is not the proper translation (Taylor, 1998, 1999). The New American Standard Version, quoted above, has the correct translation.

I also believe this upward and downward motion of the crust and mantle not only applies to the formation of mountains and valleys, but also to the large-scale movement of continents and ocean basins (Figure 7.4). There is much evidence for large-scale vertical motion. Mountains all over the Earth commonly contain fossils of marine organisms. The top of Mount Everest is a good example. It is capped by limestone with marine crinoid fossils (Gansser, 1964, p. 164), indicating that this mountain has vertically uplifted at least 29,028 feet (8850 meters) relative to sea level.

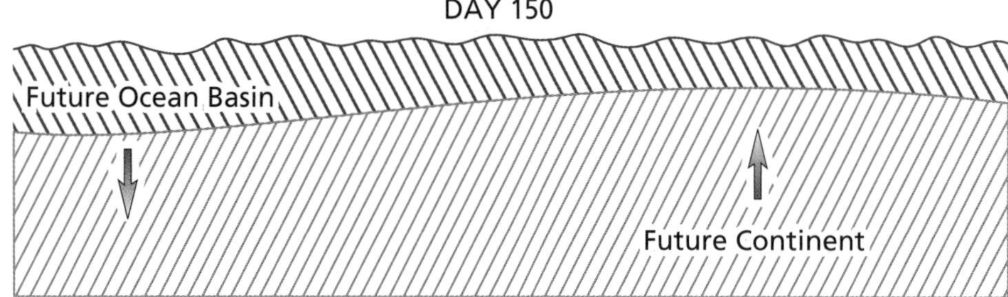

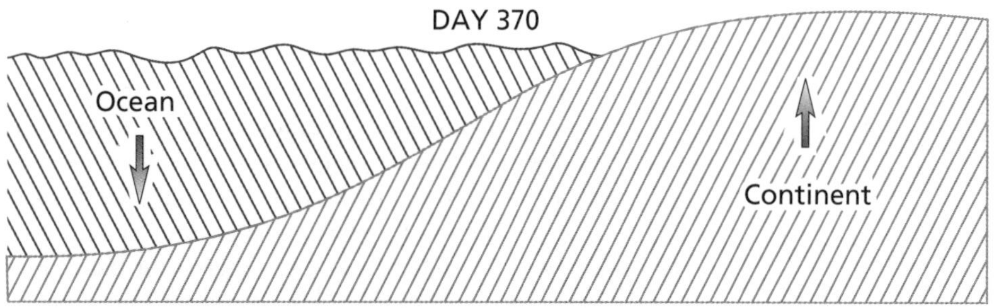

Figure 7.4 Schematic of vertical uplift of future continents and subsidence of ocean crust during the Recessional Stage of the Flood.

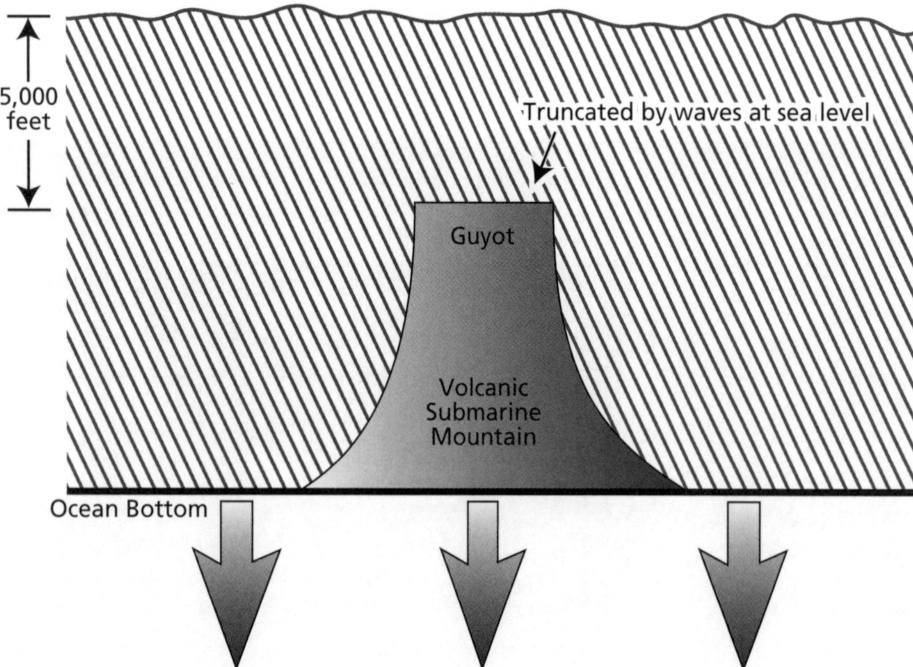

Figure 7.5 Schematic of a guyot, a volcano likely truncated at sea level producing a flat top, showing an average sinking of 5,000 feet (1525 meters) of the ocean crust. There are thousands of guyots on the ocean bottom, especially in the western Pacific.

Guyots are an indication that the ocean basins have recently undergone downward vertical motion. They are either flat-topped volcanic mountains or broken up carbonate plateaus (Figure 7.5). Thousands of guyots dot the ocean floor, especially in the western Pacific Ocean. The flat top very likely formed at sea level, but now guyots are found at about 5,000 feet (1525 meters) below sea level, although depths vary considerably (Oard, 2001a). Guyots show that large areas of the oceans have sunk around 5000 feet (1525 meters), and in *later geological time*, as stated by the uniformitarian geologist Lester King (1983, pp. 168,171):

> Marine volcanic islands which have been truncated by the waves and since subsided below sea level are called guyots. Most of them seem to have sunk by 600 to 2000 m [1970 to 6560 feet] and it is evident that they afford a measure of the amount by which the ocean floor has sunk in *later geologic time*. The Pacific floor especially has subsided...All the ocean basins afford evidence of subsidence (amounting to hundreds and even thousands of meters) in areas far from land [emphasis and brackets mine, parentheses his].

It is interesting that many critics of the Flood, both "theologians" and secular critics, have presented the "problem" that the Floodwaters had to cover the high mountains of today. There is not enough water in the oceans to cover Mount Everest, even if all water was placed at and above sea level. This is seen by some people as a contradiction to the global Flood. Bernard Ramm (1954, p. 166), a Christian theologian, wrote the very influential book, *The Christian View of Science and Scripture*, which states the problem as follows:

> If the earth were a perfect sphere so that all the waters of the ocean covered it, the depth of the ocean would be two and one-half to three miles. To cover the highest mountains would require eight times more water than we now have.

Besides his poor mathematics, Mount Everest being only about 5.5 miles (8.8 kilometers) high, the main problem with this criticism is that the mountains *rose out of* the Floodwaters; the ocean did not have to rise to cover Mount Everest. The Himalaya Mountains and most other mountain ranges rose out of the Floodwaters. Ramm (1954, p. 8) even states his unrecognized problem right in his *preface*: "With reference to technical details of the sciences I must depend on what other men say, and I am thereby at their mercy." In other words, Ramm believed anything scientists said, even about the unobservable, non-scientific past. This example is not unlike other criticisms of the global Genesis Flood.

The tremendous evidence gathered from land and sea demonstrates that the great vertical movements of the crust have occurred fairly recently. Continents and mountains have risen out of the water; continental valleys and the ocean basins have sunk. King (1983, pp. 16, 71) states that this large-scale, crustal vertical motion is *fundamental and clearly seen*:

> So the fundamental tectonic mechanisms of global geology are *vertical, up or down*: and the normal and most general tectonic structures in the crust are also vertically disposed...But one must bear in mind that every part of the globe–on the continents or in the ocean basins–provides direct geological evidence that formerly it stood at different levels, up or down, and that it is subject *in situ* to vertical displacements [emphasis his].

Figure 7.6 Devils Tower, northeast Wyoming, stands 800 feet (245 meters) above the surrounding plains and 1200 feet (400 meters) above the rivers of the region. This well-jointed igneous rock, the throat of a volcano, was once covered by sedimentary rocks. The Tower could not have remained standing for the tens of millions of years the plains were eroding all around. It is more indicative of rapid sheet erosion of the plains that left behind a few harder remnants.

Figure 7.7 Ship Rock, northwest New Mexico, stands 1700 feet (520 meters) above a wide valley. It too is an igneous erosional remnant that is the throat of a volcano, like Devils Tower.

RAPID CONTINENTAL EROSION

As the continents rise and the ocean basins sink, the water is forced to flow off the continents, sometimes at high speeds. This would result in deep erosion of the continents. Abundant evidence exists to show that the continents have been heavily eroded during the abative or sheet flow phase of the Flood. In the western United States, this is demonstrated by igneous erosional remnants that were left behind. These remnants include Devil's Tower of northeast Wyoming, 800 feet (245 meters) above the surrounding plains and 1200 feet (400 meters) above the surrounding rivers (Figure 7.6). Ship Rock, New Mexico, 1700 feet (520 meters) above a wide valley (Figure 7.7), is another example. Devils Tower and Ship Rock

Figure 7.8 Erosional remnants at Monument Valley along the border of southeast Utah and northeast Arizona.

Figure 7.9 The erosional remnant of Red Butte, near the south rim of the Grand Canyon. It is 1000 feet (300 meters) above the erosion surface at the top of the Grand Canyon.

At most a million years of erosion at the present rate would have reduced the remnants to rubble. The continents themselves can be reduced to sea level in anywhere from 10 million to about 50 million years (Roth, 1998, pp. 263-266; Schumm, 1963). Yet, Devil's Tower is believed to have remained in its general present state for well over 40 million years, despite the strong tendency for frost weathering to dislodge blocks of the strongly jointed igneous rocks.

Sedimentary erosional remnants also provide evidence for rapid erosion of at least 1000 feet (305 meters) of sedimentary rock in the western United States. They include Pumpkin Buttes in the Powder River Basin of northeast Wyoming, Tatman Mountain in the Bighorn Basin of north central Wyoming, the Cypress Hills of southeast Alberta and southwest Saskatchewan, and Monument Valley along the Arizona-Utah border (Figure 7.8) (Oard, 1996a; Oard and Klevberg, 1998). The buttes and mesas of Monument Valley consist mostly of soft sandstone that would have been reduced to a mound of rubble had they been exposed for even a million years. Red Butte near the south rim of the Grand Canyon is a sedimentary remnant capped by basalt (Figure 7.9). It is 1000 feet (305 meters) above the flat surface of the plateau (Austin, 1994b, pp. 83-84). I have previously presented a case that between 1600 to 3200 feet (500 to 1000 meters) of sediment and sedimentary rock have been eroded from the high plains of Montana, adjacent southern Canada, and northern Wyoming (Oard, 1996a, pp. 261, 262). All these remnants indicate the vast erosion of the Rocky Mountains, Basin and Range, and the high plains of the western United States has occurred recently, otherwise the remnants would not still be standing. If continental erosion developed by slow processes over millions of years as envisioned by mainstream geologists, there should be tell tale "signs." The following scenario is one of the sorts of "signs" we can imagine. The eroded material should have formed thick deposits of alluvium and sediments with terraces from

(Plummer and McGeary, 1996. pp. 77-78) are very likely the throats of eroded volcanoes. All of these igneous rocks had to *intrude* into other sedimentary rocks, which no longer exist. It is easy to conclude that sedimentary rocks once surrounded and covered these igneous remnants. The remnants were a little more resistant to erosion than the surrounding sedimentary rocks.

These remnants provide evidence for very rapid erosion of the surrounding sedimentary rock. They could not be standing today if the surrounding rocks eroded slowly over millions of years, as believed by uniformitarian scientists.

the highest land all the way to the coast, especially along river and stream valleys. The continents should be one large waste surface inclined towards the coasts. Once the debris entered the coastal zone, it should have fanned out from the river mouths as huge deltas. Currents and slides would rework some of the sediment along the continental shelf and into the deep ocean. Thousands of feet of rock have been removed from the continents. Where has all the material gone? The debris from postulated slow erosion over millions of years is not what we observe on the continents. Although we do have river deltas, they are not even close to large enough for the massive continental erosion that has occurred. The only conclusion left is that the debris has been totally removed from the continents by some incredible mechanism.

RAPID FORMATION OF CONTINENTAL SHELVES BY SHEET DEPOSITION

In the Flood model being presented here, large-scale sheet erosion would deposit sediments in areas of waning currents. A good analogy would be how the Portland Delta formed during the Lake Missoula flood (Bretz, 1928d, pp. 697-700). The floodwaters rushed through the Columbia Gorge at more than 80 mph (35 m/sec), then slowed when they came to the wide mouth of the gorge in the Portland, Oregon and Vancouver, Washington, area. The waning current deposited a giant sand and gravel delta up to 350 feet (110 meters) thick and 200 mi^2 (500 km^2) in area.

Sheet flow from the rising continents during the Genesis flood would drop their load of sediments in areas where the currents slow, which would be at the edge of deep water. This area would be along the continental margins or in large, low elevation continental basins at the edge of the continents, such as the Lower Mississippi River Valley (before being filled by sediments that have since hardened). The eroded material from the continents would then form the continental shelves, slopes, and rises. Much finer-grained sediment would be transported farther out into the ocean and form some of the clay deposits of the abyssal plains. Turbidity currents that move at high speed down steep slopes probably were far reaching during the Recessional Stage and deposited sand far from the continents into the abyssal plains of the deep ocean.

Continental shelves are enigmatic from a uniformitarian point of view. Figure 7.10 is an illustration of the continental margin showing the continental shelf, slope, rise, and abyssal plain. The shelf is very flat with a slope of less than 0.1 degree and a relief of less than 65 feet (20 meters) (Kennett, 1982, p. 29). It averages 50 miles (78 kilometers) wide but varies from a few miles (several kilometers) to over 300 miles (480 kilometers). The Bering Sea shelf and the Grand Banks are good examples of wide continental shelves. Then, suddenly, at the shelf break at an average depth of 425 feet (130 meters), the slope abruptly becomes significantly greater (approximately 4^0). This narrow zone plunges to depths of 5000 feet to 11,500 feet (1500 to 3500 meters) below sea level to the 60-600 mile (100-1000 kilometer) wide continental rise. The rise then slopes gently downward to an abyssal plain. These features of the continental margin are difficult to explain by mainstream geologists because natural processes would favor a gradual descent from the coast to the ocean depths; there should be no continental shelf or slope. Seismic reflection profiles generally show broad delta-like features prograding seaward as well as very gentle oceanward-dipping sediments with the slope of the sediments greater the deeper the sediment. Lester King (1983, pp. 199-200) describes continental shelves, and the problems they present to uniformitarianism:

> There arises, however, the question as to what marine agency was responsible for the leveling of the shelf in early Cenozoic time, a leveling that was preserved, with minor modification, until the offshore canyon cutting of Quaternary time? Briefly the shelf is too wide, and towards the outer edge too deep, to have been controlled by normal wind-generated waves of the ocean surface...The formations and unconformities have been tilted seaward (monoclinally) at intervals during the later Cenozoic. There have been repeated tectonic episodes: always in the same sense - *the lands go up and the sea floor down...* [emphasis mine].

In his book, *The Natal Monocline*, King (1982, p. 45) further adds in reference to the continental shelf:

> We note that all the formations drilled dip offshore. The oldest and deepest formations dip at several degrees, the youngest and uppermost dip at less than one degree.

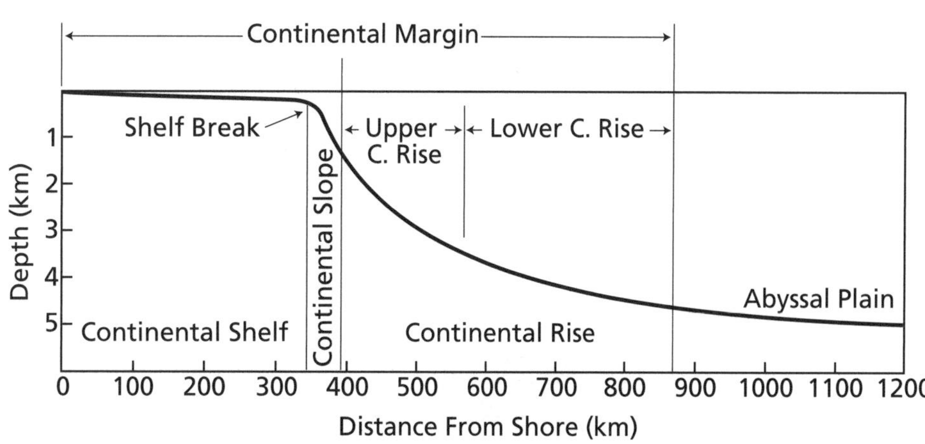

Figure 7.10 Principal features of the continental margin with a vertical exaggeration of 50 times (from Kennett, 1982, p27; redrawn by Nathan Oard)

The uplift of the continents and sinking of the ocean basins echoes Psalm 104 quoted previously. The strata of the continental shelves are quite flat and extend far off the coast as a sheet of sedimentary rock, which is unexpected if the earth were millions of years old. Over such a long period of time, there should be a steady descent of the depth of the continental margin into the abyssal plains, not the nearly flat bottom, also reflected in the strata below, of the continental shelf, followed by a "drop off" at the continental slope. It is interesting to note that the sediments of the shelf are not only remarkably flat, but also that very few canyons have been cut into them–until after nearly all the sediments were deposited as a sheet. This is a similar situation as on the continents. Fulthorpe and Austin (1998, p. 262) write in regard to the "Miocene" age of the sedimentary rocks of the United States Atlantic shelf: "The rarity of middle-upper Miocene clinoform slope canyons contrasts starkly with conditions on the heavily dissected modern slope." This lack of canyons cut in the planar beds of the continental margins until after nearly all the sediments were laid down is why King refers to the offshore canyon cutting phase as occurring during the *Quaternary*, the last period of geological time. Within the Flood model extensive submarine canyons, many deeper and at least one longer than the Grand Canyon, were cut *after* sheet deposition took place during the dispersive (channelized) phase of the Flood. This is when accelerated, erosive channelized currents descended off the continents onto the newly deposited continental shelf, eroding deep, channelized gouges. Both the continental margins and the submarine canyons are difficult to explain by applying uniformitarianism, yet expected in the Flood model.

PLANATION SURFACES AND LONG-TRANSPORTED BOULDERS

When the earth was completely flooded, a number of mechanisms would have generated powerful currents during the Inundatory Stage of the Genesis Flood (Oard, 2001a). One of these is the Coriolis force caused by the spin of the earth, which can generate currents from rest to over 100 mph in about one month on large land masses covered by less than 3300 feet (1000 meters) of water (Barnette and Baumgardner, 1994). During the start of the abative phase of the Recessional Stage, these wide currents would continue to sweep as a sheet over the relatively shallow continents. As the future continents continued to uplift, the Floodwater was forced to flow off of the continents toward the future ocean basins, sometimes at very high speeds (Figure 7.11). Over time as the land gradually emerged more and more, the currents would be forced to become narrow and channelized. Linear erosion would accelerate. Along the way the water picked up a tremendous amount of sediment, especially from the elevated areas of the future continents.

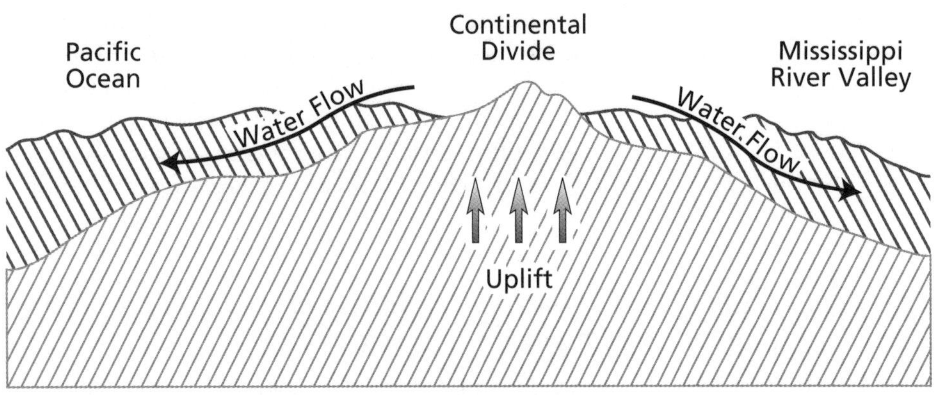

Figure 7.11 Schematic of Flood waters rushing off the rising continents (Drawn by Peter Klevberg and modified by Daniel Lewis).

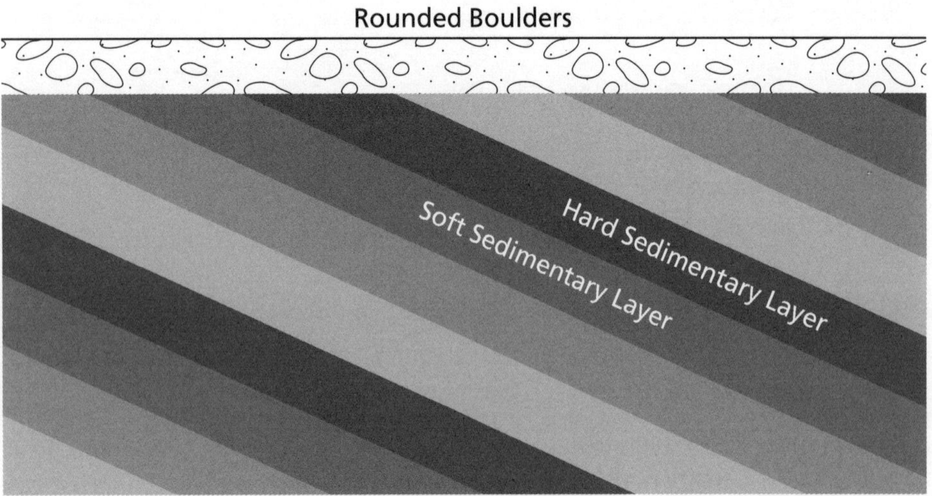

Figure 7.12 Schematic of a gravel-capped planation surface truncating both tilted hard and soft rocks the same (Drawn by Peter Klevberg and modified by Daniel Lewis).

The Recessional Stage was the last event of the Flood. After this stage was completed most areas of the earth have experienced very little erosion. So, the topography we observe today should give us copious evidence for this last great event of the Genesis Flood–the runoff of the water. In the study of landforms, the province of geological discipline of geomorphology, we observe what we expect to find from the Flood paradigm. Evidence of sheet flow having scoured the continents is largely provided by remnants of large flattened planation or erosion surfaces. A planation or erosion surface is defined in the *Dictionary of Geological Terms* (Bates and Jackson, 1984, p. 170) as: "A land surface shaped and subdued by the action of erosion, especially by running water. The term is generally applied to a level or nearly level surface." Planation surfaces are not plains of deposition, like a flood plain, alluvial fan, or river terrace. Planation surfaces are *smoothly planed rock*. The difference between a planation surface and an erosion surface is that an erosion surface can be more rolling. The rocks cut by the planation or erosion surface can be either hard, soft, or a combination of both. It does not matter. Hard and soft rocks commonly are both smoothed the *same amount* by the eroding mechanism. A planation surface can be eroded on horizontal sedimentary rocks, but the most significant and easily identified planation surfaces are cut on tilted sediment rocks. Figure 7.12 illustrates a planation surface cut across interbedded, tilted hard and soft sedimentary rocks that are covered with a veneer of rounded cobbles and boulders left over from the aqueous agent that eroded the surface. These types of planation surfaces are commonly observed.

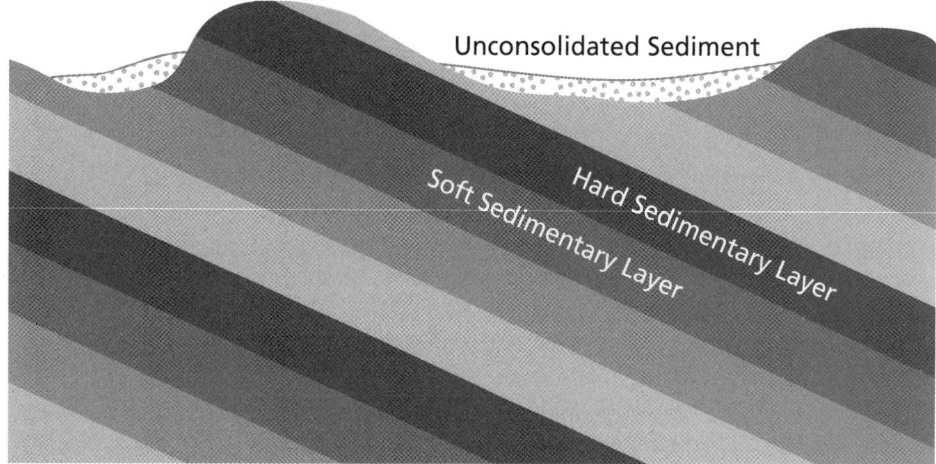

Figure 7.13 Schematic of what would be expected from the uniformitarian model of slow erosion over millions of years. Soft rocks would be more eroded than hard rocks, leaving hard rocks as ridges (Drawn by Peter Klevberg and modified by Daniel Lewis).

Figure 7.14 A flat gravel-capped planation surface along the southeast edge of the Little Rocky Mountains, north central Montana. Note that the planation surface is being dissected by streams issuing from the mountains (behind photo).

After planation surfaces formed, subsequent Flood processes modified the surfaces. Continued Flood erosion would carve away at the edge or dissect a newly formed planation surface. Thus, only remnants of the original surface are all that remain. The planation surface can also be completely obliterated, leaving behind only mountains with accordant summits. Continued vertical tectonics could cause a planation surface to fault with parts ending up at different elevations. In other situations, volcanism can sometimes cover a planation surface. Thus, some planation surfaces can be more difficult to recognize.

Assuming the uniformitarian paradigm that projects normal erosion occurring over long periods of time, the softer rocks of an area should have eroded faster than the harder rocks. If a generally flat surface somehow formed, it would be significantly roughened with time, much like the schematic shown in Figure 7.13. We do not observe planation surfaces being formed today, except rarely on a very small scale when a flooded river suddenly shifts and cuts its bank. Instead, we observe the world's existing planation surfaces being eroded and dissected (Figure 7.14). Present processes do not favor their formation or their preservation. For these two reasons planation surfaces are relics of a past process and not a result of present processes. Furthermore, because generally flat surfaces would be roughened with time, and because even the world's planation surfaces cut entirely on soft rocks have not been eroded away, planation surfaces are relatively *young* landforms, most probably thousands not millions of years old (see Chapter 8).

In the erosional process of the Flood, soft rocks such as shale would be eroded and quickly pulverized by the draining Floodwaters. More resistant lithologies, such as quartzite

Figure 7.15 Gravel-capped planation surface of central block of Cypress Hills, southeast Alberta, Canada.

Figure 7.16 Conglomerate cap of Cypress Hills planation surface composed of well-rounded and iron-stained quartzite gravel, cobbles, and boulders.

and chert, would break apart as blocks and become smaller and more rounded as they were transported long distances by the Flood waters. These rocks eventually would disintegrate as well. Therefore, during the Flood's rapid sheet erosion, resistant rocks would be transported far from their sources. We would expect to find these rounded, resistant rocks often on top of the planation or erosion surfaces, as well as in other areas. We observe over eastern Montana and adjacent Canada that erosion has left behind generally four planation surfaces at different altitudes (Alden, 1932). Peter Klevberg and I have been studying the highest two surfaces: the Cypress Hills of southeastern Alberta and southwest Saskatchewan (Figures 7.15 and 7.16) and the Flaxville Plateaus of northeast Montana (Klevberg and Oard, 1998; Oard and Klevberg, 1998). The Cypress Hills are a high, flat plateau about 1000 feet (305 meters) above the next highest planation surface and about 2500 feet (760 meters) above the rivers to the north and south. Therefore, at least 2000 feet (760 meters) of sedimentary rock has been eroded from the region around the Cypress Hills. The Cypress Hills planation surface had an area of about 780 mi^2 (2000 km^2), before being slightly dissected, probably by glaciofluvial activity at the end of the ice age.

The boulders and cobbles on both the Cypress Hills and Flaxville Plateaus are well-rounded and mostly iron-stained quartzite rocks with abundant percussion marks (Figure 7.17). The percussion marks indicate that the rocks had slammed hard against each other. The cobble and boulder cap is mostly clast supported. The data indicate that the cobbles and boulders were deposited by water and not mud flows, although mudflows could have transported the material at times. This massive layer of boulder and cobbles caps approximately 80 feet (25 meters) of the top of Cypress Hills plateau. There are only a few sand interbeds in the western and central blocks of the Cypress Hills (Vonhof, 1965). The largest rock we observed was in the western Cypress Hills and has a length of 15 inches (39 centimeters) and a width of 9 inches (24 centimeters) with a weight of 58 pounds (26 kilograms). Based on inferred west southwest paleocurrent directions, the nearest source for the quartzite rocks is the Rocky Mountains of northwest Montana (Vonhof, 1965). Thus, the quartzite rocks have been transported over a slope ranging from less than 0.1 degree to a maximum of 0.4 degree (Klevberg and Oard, 1998, p. 372) for a distance of 190 miles (300 kilometers) to the western Cypress Hills and 450 miles (700 kilometers) to the eastern Flaxville Plateau. Some researchers suggest that the coarse gravel may have originated from central Idaho (Leckie and Cheel, 1989). If this is the case, we have to add another 125 miles (200 kilometers) to the above distances. We have also found these rocks as far away as western North Dakota and southeast Montana, so we could add even another 100 miles (160 kilometers) of transport. These exotic quartzite rocks have been transport on a low slope up to *660 miles (1060 kilometers)*!

Intuitively we know that modern rivers cannot transport such huge amounts of cobbles and boulders anywhere near this distance, especially on such low slopes. To estimate quantitative current velocities, current depths, and other paleohydrological variables, Klevberg employed standard coarse-sediment hydrologic equations. He calculated that to transport the boulders as bedload to the Cypress Hills, minimum current velocities of 9 to 13 mph (4 to 6 m/sec) with minimum water depths of 10 to 130 feet (3 to 40 meters) are required over a broad area. Unless very narrow channels are postulated, for which there is no evidence, the resulting discharges had to have been hundreds of times greater than any historic regional flood.

Furthermore, the abundant percussion marks imply that many of the pebbles and cobbles were transported in *intermittent suspension*. Based on the size and shape of these rocks, and the relationship between the current velocity and the fall velocity during suspension (Blatt, Middleton, and Murray, 1972), Klevberg calculated a *minimum current velocity of 65 mph (30 m/sec)* with a flow depth of at least *180 feet (55 m)* (Klevberg and Oard, 1998). The geomorphology of the flat-topped Cypress Hills alone indicates a sheet flow of this size *at least 12 miles (20 kilometers) wide*. There is evidence the Cypress Hills planation surface, when first formed, extended north and south a considerable distance (Alden, 1932). Thus such a current would probably be hundreds of miles (many hundreds of kilometers) wide. The current necessary to produce the sheet of gravel capping the Cypress Hills and Flaxville Plateaus in both lateral and run-out distance exceeds any conceivable flash flood by hundreds or thousands of times. These coarse gravel caps are not the result of a glacial Lake Missoula-like jökulhlaup (burst glacial lake) because there are no glacial erratics or glacial debris in the coarse gravel. The rocks have no relationship to glaciation; they are pre-glacial. The source of the rocks is the Rocky Mountains, probably from west of the continental divide based on the degree of metamorphism of the quartzite rocks. We have concluded that the Cypress Hills and Flaxville Plateaus boulder- and gravel-capped planation surfaces defy uniformitarian pro-

Figure 7.17 Large percussion marks with a diameter of about 4 inches (10 centimeters) from the Cypress Hills cobble and boulder cap, southeast Alberta. Practically all percussion marks on the Cypress Hills rocks are only about 1 inch (2.5 centimeters) in diameter. Percussion marks on well-rounded hard rocks indicate violent transport in water.

Figure 7.18 Flat-topped granite mountains, northwest Wind River Mountains, Wyoming.

cesses and are more consistent with the Genesis Flood (Oard and Klevberg, 1998; Oard, 2000b).

Planation surfaces, occasionally capped with well-rounded cobbles and boulders, are observed regionally, in the northwest states. Planation surfaces are most impressive when they are at the *tops* of mountains. Several examples of this can be found at about 11,500 feet (3,500 meters) ASL in the northwest Wind River Mountains, Wyoming (Figures 7.18 and 7.19) and nearly 13,000 feet (4000 meters) ASL in the Beartooth Mountains of south central Montana (Figures 7.20 and 7.21). Well-rounded quartzite rocks from the Rocky Mountains of western Montana and northern and central Idaho are found not only on the Cypress Hills and Flaxville Plateaus, but also can be found, sometimes in very thick deposits, in southwest Montana, southeast Idaho, western Wyoming, northern and central Oregon, southern Washington, and as far west as Astoria and Hoquiam on the Pacific Ocean. Well-rounded quartzite cobbles are commonly found in the Puget Sound area, even on the San Juan Islands, mixed in, as a minor component, with other glacial lithologies (Mustoe, 2001). These quartzite rocks were most likely brought into the Puget Sound area or southwest British Columbia by flows of water and then reworked by the glaciers (Mustoe, 2001, p. 19). In checking Mustoe's report, I needed to examine just one outcrop of glacial debris in the Puget Sound area. In a section along a creek about 25 miles (40 kilometers) northeast of Everett, Washington, I found that approximately one out of 200 rocks was a well-rounded quartzite cobble. It is interesting that a few cobbles even possessed abundant percussion marks indicating catastrophic flows of water that emplaced the cobbles. The Puget Sound lobe of the Cordilleran Ice Sheet has barely eroded the surface of these cobbles, indicating glaciation was not very intense and long. One of the most interesting locations for quartzite boulders is on ridges in the Wallowa Mountains of northeast Oregon. Figure 7.22 shows a 30-foot (10 meters) thick outcrop of well-rounded quartzite boulders up to 3 feet (1 meter) in diameter on a ridge just southeast of Lookout Mountain at 8200 feet (2500 m) ASL. The largest boulder we found was 2 feet (0.7 meter) in diameter with an estimated weight of about 440 pounds (200 kilograms) (Figure 7.23). The nearest outcrop of quartzite is at least 60 miles (100 kilometers) east in central Idaho. John Eliot Allen, late professor of geology at Portland State University, discovered these quartzite boulders on eight ridges in 1938. He declared that the boulders haunted him ever since (Allen, 1991, p. 104). It troubled him to think they must have been deposited by some kind of *torren-*

Figure 7.19 Flat-topped Gypsum Mountain, northwest Wind River Mountains, Wyoming. This mountain is composed of limestone dipping about 30° to the west (right). Gypsum Mountain is located just west of the granitic mountains shown in Figure 7.18, indicating planation sheared both igneous and sedimentary rocks in the area.

Figure 7.20 The flat-topped highest peaks of the Beartooth Mountains, south central Montana. This planation surface is cut in granite.

Figure 7.21 Close up of Beartooth Mountains planation surface. Note that the planation surfaces that shear the top of the granite peaks are at different elevations, probably from slight faulting.

Figure 7.22 Outcrop of large quartzite boulders on a ridge near Lookout Mountain at 8,200 feet (2,500 meters) in the Wallowa Mountains, northeast Oregon.

tial paleocurrent, as he called them. Recently, field trips by the Design Science Association of Portland, Oregon, led by John Hergenrather have discovered similar but more extensive quartzite cobbles and boulders scattered throughout the western Wallowa Mountains. All of these many outcrops of quartzite cobbles and boulders indicate that large-scale erosion and long distance transport both east and west of the Rocky Mountains occurred during a period of massive regional erosion. Since the boulders are sometimes found one to several mountain ranges away from their nearest source, their transport implies that the uplift of the mountains had not yet proceeded or was just beginning while the boulders were being transported many hundreds of miles. Mountain top erosion surfaces were formed during the Flood while the northwest states were generally a flat, planed surface. Uniformitarian geologists, Ollier and Pain (2000), have documented that before the final mountain uplift, all the continents were planed. The rugged relief of the Pacific Northwest, therefore, was produced during the Recessional Stage of the Flood due to vertical tectonics (valleys and canyons would be cut in the dispersive or channelized phase of the Flood).

Moreover, planation surfaces, some of extraordinary smoothness, are not only found in the Northwest United States, but they are also a *worldwide* phenomenon. Lester King (1967) documents this in his book, *The Morphology of the Earth*. Sixty percent of Africa is a planed erosion surface at one or more levels (King, 1967, pp. 241-309). In case anyone thinks King's work has been refuted since he published it, well-known Australian geomorphologist, C.R. Twidale (1998, p. 660) in a recent and provocative article on supposed very old erosion surfaces admits that King's grand scheme of multiple worldwide planation surfaces is *generally correct*. Twidale shows that within the uniformitarian time scale, many of these erosion or planation surfaces have remained nearly flat for many tens of millions of years. He is attempting to find a mechanism that would protect the surfaces from erosion over that vast time (see Chapter 8).

Many hypotheses have been attempted by mainstream geologists to account for these large-scale surfaces, but all of them have serious difficulties. Crickmay (1974, pp. 192, 201) summarized the deplorable state of geomorphic research on landforms in 1974, including hypotheses for the formation of planation surfaces:

> The difficulty that now confronts the student is that, though there are plenty of hypotheses of geomorphic evolution, there is not one that would not be rejected by any majority vote for all competent minds.

Figure 7.23 Two-foot well-rounded quartzite boulder on the ridge shown in figure 7.22. (photo by Paul Kollas).

Figure 7.24 Susquehanna water gap through the Appalachian Mountains as seen from Harrisburg, Pennsylvania. The Susquehanna River generally cut straight through 4 or more ridges but only the gap through the last ridge can be seen due to a broad meander in the river.

> This situation is in itself remarkable in a respectable department of science in the latter half of the 20th Century...A century and a half of literature bearing on scenery and its meaning shows primarily the inspired innovations that carried understanding forward; followed in every case by diversion from sound thinking into inaccuracy and error.

I may add that the situation is little different today. Geomorphologists have mostly given up trying to explain large-scale landforms and during the past 40 years have focused on what is called process geomorphology, which deals with small-scale present processes that are shaping the land, such as river erosion.

Planation surfaces are a powerful witness to the veracity of a global Flood and a strong witness against slow processes forming them over millions of years. Planation surfaces are a documentation of sheet erosion occurring during the abative phase of the Recessional Stage of the global Flood, a Flood that left planation surfaces on *all of the continents.*

WATER AND WIND GAPS FORMED RAPIDLY DURING THE CHANNELIZED PHASE OF THE FLOOD

Water and wind gaps are other landforms on the surface of the Earth that geologists find difficult to explain using uniformitarianism. A water gap is: "A deep pass in a mountain ridge, through which a stream flows; esp. a narrow gorge or ravine cut through resistant rocks..." (Bates and Jackson, 1984, p. 559). A wind gap is defined as: "A shallow notch in the crest or upper part of a mountain ridge, usually at a higher level than a water gap" (Bates and Jackson, 1984, p. 564). A wind gap is considered an ancient water gap that was subsequently abandoned as erosion continued to lower the valleys. Many classic examples of wind and water gaps can be found in the Appalachian Mountains, such as along the Susquehanna River (Figure 7.24). Early publications are replete with references to this puzzle (Ver Steeg, 1930). Williams *et al.* (1994) analyzed the Pine Creek gorge water gap in Pennsylvania. Water gaps are only impressive if the stream could have chosen an easier path around a rock barrier *instead of appearing to cut right through it.* Only this type is of interest to us. Unaweep Canyon, a narrow canyon cut about half way down through the Uncompahgre Mountains of western Colorado (Figure 7.25), is an impressive example of a wind gap (Shaver, 1998; Oard, 1998c; Williams, 1999).

There are well over one thousand water gaps located all over the earth. Many of these have incongruously deep gorges. For example, the Shoshone River west of Cody, Wyoming, cut a water gap about 2000 feet (600 meters) deep through the granite-cored Rattlesnake Mountains (Figures 7.26). The river could have easily gone around the Rattlesnake Mountains a few miles (several kilometers) to the south when the valley was at a slightly higher level. Surprisingly, there is not much difference between the altitude of the plains and this possible route to the south (Figure 7.27).

Another example is the Yakima River that flows through Ellensburg to Yakima, Washington (Oard, 1996a, pp. 270-271). The Yakima River could have easily kept flowing east from Ellensburg into the Columbia River, but instead it took an abrupt southward turn and cut incised meanders through at least four anticlines of the Columbia River Basalt Group.

Hells Canyon is a water gap 50 miles (80 kilometers) long and up to 8000 feet (2400 meters) deep that has been cut through the Wallowa Mountains of northeast Oregon and the Seven Devils and Cuddy Mountains of west central Idaho (Vallier, 1998). The Snake River could have more easily flowed west from southern Idaho into southeast Oregon. Supposedly this water gap is only 2 to 6 million years old in the uniformitarian time scale, and yet there is little if any trace of the former river in Oregon.

The Sweetwater River of Wyoming transects the nose of an exhumed, plunging anticline when it could have easily flowed around the barrier only half a mile (one kilometer) to the south (Thornbury, 1965, p. 359) (Figure 7.28).

Eleven rivers develop on the southern Tibetan Plateau or the north slopes of the Himalaya Mountains and cut through the full width of the range through deep gorges (Oberlander, 1985). One of these rivers, the Arun River, has cut a few tens of thousands of feet (over 6 kilometers) through a transverse anticline east of Mount Everest!

The Zagros Mountains, southwest Iran, have peaks commonly in the 11,000 to 15,000 feet (3,350 to 4,600 meter) range with more than 300 water gaps (Oberlander, 1965). The deepest one is about 8000 feet (2400 meters) deep, deeper than the Grand Canyon. These water gaps, cut through mountains that rose in the "Pliocene" and "Pleistocene" of the uniformitarian geological time scale, seem to defy rationality. Here is a brief sampling of Oberlander's (1965, pp. 1, 16, 21, 89) description of the amazing Zagros water gaps:

> The Zagros drainage pattern is distinctive by virtue

Figure 7.25 Unaweep Canyon wind gap through the Uncompahgre Plateau of western Colorado.

Figure 7.26 The Shoshone water gap cut about 2000 feet of granite and sedimentary rocks through the southern end of the Rattlesnake Mountains, west of Cody, Wyoming.

of its *disregard* of major geological obstructions, both on a general scale and in detail...[In the central Zagros] major streams utilize longitudinal valleys to a *minimum* degree, despite the presence of the greatest structural barriers to be found in the orogenic system...In a surprising number of instances plunging fold noses are crossed by engorged transverse streams although open valley paths pass the ends of the ridges less than a mile away...Certain streams ignore structure *completely*; some appear to "seek" obstacles to transect [emphasis mine, quotes his].

There are several instances of a stream cutting through the *same* transverse ridge anywhere from two to five times. This would be equivalent to the Willamette River of western Oregon cutting through the Cascade Mountains to the east and then back again–twice! The Zagros drainage system is distinctive, but similar water gaps are found in other mountain ranges:

> The drainage history of this region is as obscure as is that of most of the Cenozoic and older mountain systems of the world whose transverse streams have been deduced, in the absence of evidence to the contrary, to be antecedent, superimposed, or the result of headward extension under unspecified controls (Oberlander, 1965, p. 149).

Figure 7.27 Buffalo Bill Reservoir and low area west of the Rattlesnake Mountains in which the Shoshone River could have flowed south (view into the picture) around the mountains when the river ran higher in the past.

Figure 7.28 A water gap on the Sweetwater River called the Devil's Gate, cut in granite 350 feet (107 meters) deep. The river could have easily passed to the south half a mile (one kilometer) when the valley sediments were higher.

There are three major uniformitarian hypotheses to account for water gaps, as mentioned above by Oberlander: antecedence, superposition, and piracy (Williams, Meyer, and Wolfrom, 1991, 1992a,b; Austin, 1994a, pp. 85-92). Antecedence is the idea that a stream continued to flow and erode over the same location while a mountain barrier slowly lifted upward across its path. Superposition is the hypothesis that the drainage pattern was cut while the whole land was nearly flat. The rivers cut perpendicular down through hard rock that was later left standing while the softer rocks, now the valleys, were eroded. Stream piracy is the idea that one stream erodes faster than another and eventually captures the headwaters of the second stream. However, there is little evidence that any of these hypotheses could have effectively accomplished the work:

> Large streams transverse to deformational structures are *conspicuous* geomorphic elements in orogens of *all ages*. Each such stream and each breached structure presents a geomorphic problem. However, the *apparent absence of empirical evidence* for the origin of such drainage generally limits comment upon it. Transverse streams in areas of Cenozoic deformation

> are routinely attributed to stream antecedence... where older structures are involved the choice includes antecedence, stream superposition from an unidentified covermass, or headward stream extension in some unspecified manner. Whatever the choice, *we are rarely provided with conclusive supporting arguments* [emphasis mine] (Oberlander, 1985, pp. 155, 156).

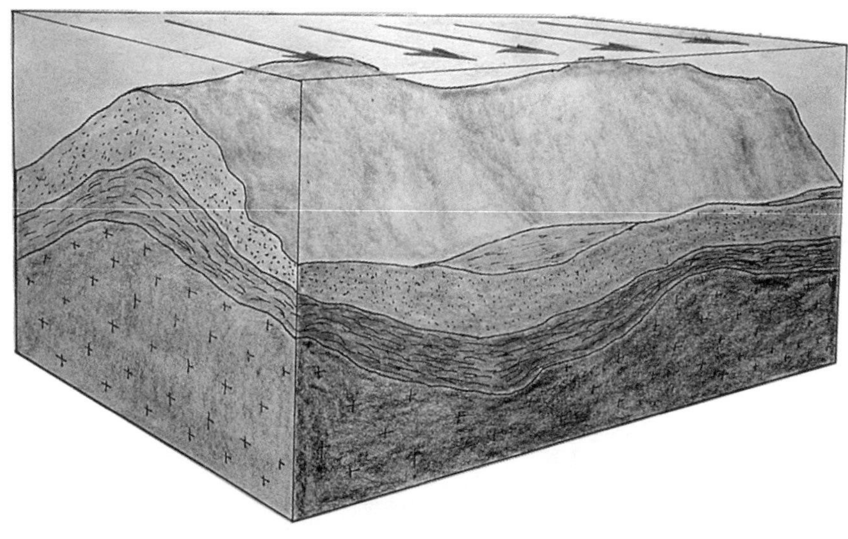

Figure 7.29 Series of schematics on the formation of water and wind gaps (drawn by Peter Klevberg). a) Water flowing perpendicular to a transverse ridge forms shallow notches on the ridge.

Notice that water gaps are conspicuous features in mountains of all ages and that there rarely is any concrete evidence to support any of the three hypotheses. Therefore, water gaps are:

> ...one of the more perplexing and ubiquitous enigmas of regional physiography; the anomaly of through-flowing drainage that is transverse to the structure of an orogenic system (Oberlander, 1965, p. 1).

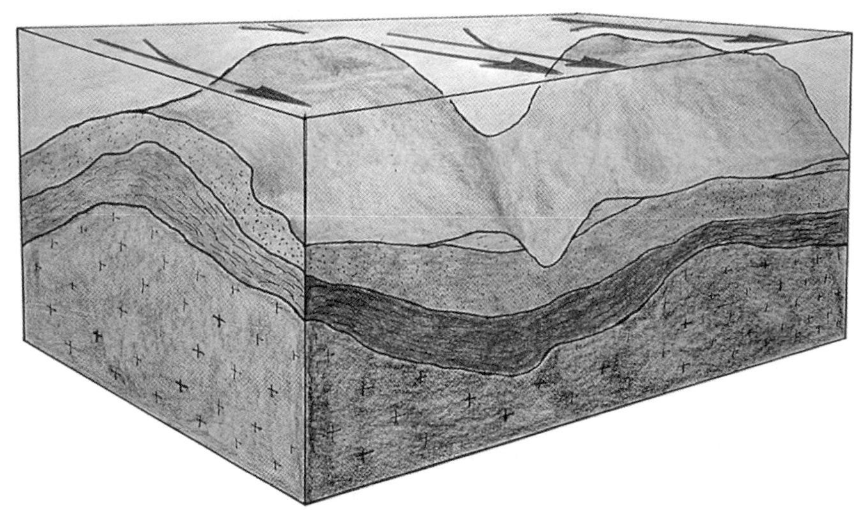

Fig 7.29. b) Notches eroded further as the water drops below the top of the ridge.

Water, as well as wind gaps, could have been formed rapidly and easily during the massive erosion that occurred during the Recessive Stage of the Flood, especially during the dispersive phase. The gaps could have been cut by Flood currents flowing transverse to the structure either while there was a "covermass" over the structure or after the ridge became more exposed. In chapter 9, I will show how the Lake Missoula flood provides an analog for the rapidly cutting of water and wind gaps through a transverse ridge. The initial notch in a transverse ridge could have been initiated during the abative or sheet flow phase of the Flood. It is not unusual for a sheet flow to have areas of enhanced flow (Schumm and Ethridge, 1994, p. 11), which would cause greater erosion and initiate a notch in a barrier ridge. Another way a notch could initially be eroded is if a portion of the ridge possessed less consolidated sediments or was locally faulted. Once the notch first forms, the water would tend to funnel with higher velocity through the notch. The abrading material being carried along the bottom of the flow would quickly cut out the water or wind gap. Post-Flood erosion, especially associated with the ice age, could have further cut the gap. Emmett Williams (1998) postulated that the Black Canyon, Colorado, water gap formed during the late Flood channelized erosion and was followed by further erosion during high water from the ice age. Figure 7.29 presents a series of schematics on how I visualize the formation of water and wind gaps during the Flood. The dispersive or channelized phase of the Recessive Stage of the Flood can explain these mysterious geomorphic features that are found worldwide. Stated another way, water and wind gaps provide strong evidence for the channelized phase of the Genesis Flood.

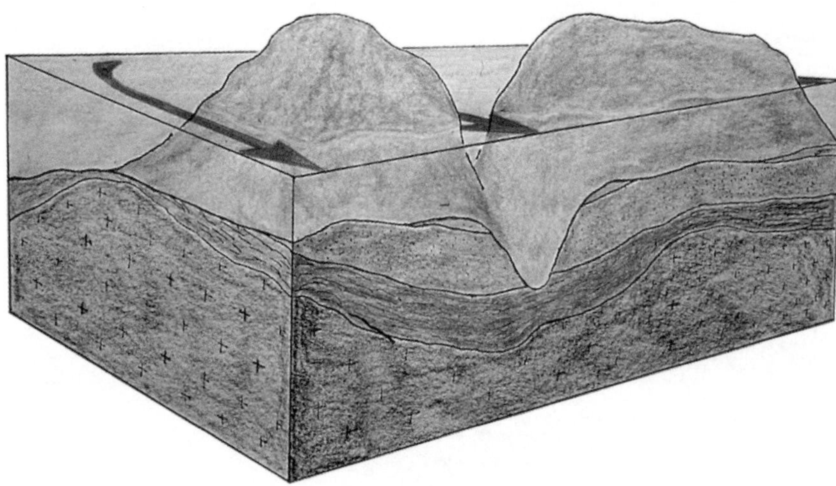

Figure 7.29. c) Floodwaters continue to drain as notches deepen.

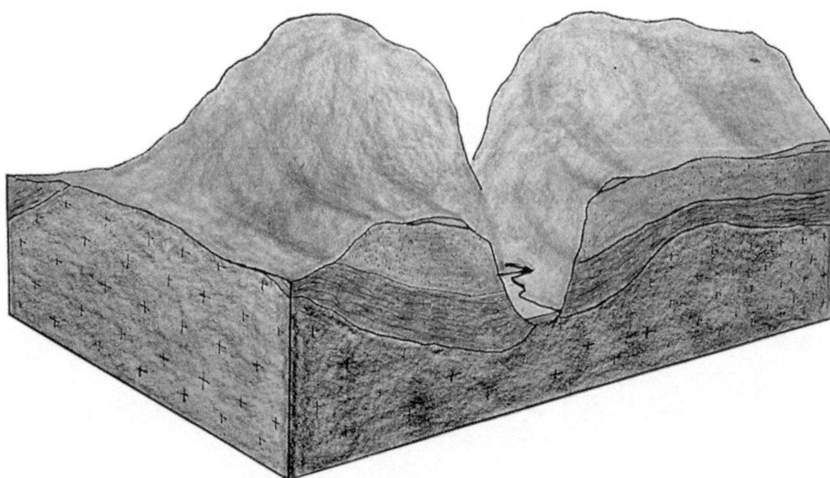

Figure 7.29. d) Floodwaters completely drained with a river running through the lowest area of the water gap. Erosion ceased too early through the other notch, leaving a wind gap.

Chapter 8

The Question of Time

Vast layers of sedimentary rocks, their contained fossils, and unique features of the Earth's surface provide compelling evidence for a global Flood as described in the book of Genesis. If the fossils are really a result of the Flood, they cannot represent an evolutionary sequence. They are a burial sequence during the Flood. The universal gaps in the fossil record, especially evident at the higher biological classification categories (Gish, 1995), indicate that evolution never happened. This all lends tremendous support for the early chapters of Genesis. Chapters 5, 10 and 11 –the genealogies –teach that the Earth is young, around 6000 years old or slightly older. In this day and age when we constantly hear of millions and billions of years for the ages of rocks and fossils, a young age from the Bible appears to be preposterous, or is it?

The "last" Lake Missoula flood within the uniformitarian time scale is dated at about 13,000 years ago (Mullineaux *et al.*, 1978). This date is an example of a date older than the Earth according to Scripture. In the Scriptural time scale, the Lake Missoula flood occurs after the Flood at the end of the ice age (Oard, 1990) and would be dated about 4000 years ago. If the lake Missoula flood occurred only 4000 years ago, what about the older uniformitarian date? What about all those radiometric dating methods that give dates millions and billions of years old for rocks? We are led to believe that radiometric dates are based on solid science; they are "absolute." Doesn't it take millions of years for many features of the rocks to form, such as coal and oil? Since the Flood is supposed to account for practically all of the sedimentary rocks and fossils, what about all those features of the rocks and fossils that appear to take too much time for a one year Flood? Two examples that come to mind are dinosaur footprints and eggs, which must have formed during the Flood, since footprints and eggs are found *within* sedimentary rock. Both radiometric methods and the many processes believed to take too much time for the Scriptural time scale challenge not only the Flood explanation for rocks and fossils, but also the straightforward meaning of Scripture.

The problem of millions and billions of years, probably more than any other idea, keeps people from believing in creation and the Flood from the early chapters of Genesis. It has also been one of the main sources of doubt that has shaken faith in the Bible as God's truthful communication to us. There actually is a good defense for Scripture and its time scheme, if we accept there was a global Flood.

The same bias that resulted in rejection of the Lake Missoula flood for 40 years operates to keep the millions and billions of years flowing out to the public, despite the fact that these dating methods rely on many assumptions and have serious problems. Much recent research by both creationists and evolutionists has opened up some unseemly cracks in the radiometric dating systems. Other discoveries prove that many processes thought to take millions of years can be accomplished rapidly under the proper conditions. For instance, it has been shown in the laboratory that coal and oil can form in a very short time under the right conditions –millions of years are not necessary (Hill, 1972; Bruce *et al.*, 1996).

SCHOLARS HAVE ASSUMED AN OLD EARTH SINCE THE SEVENTEENTH CENTURY

It may come as a surprise to learn that the concept of an old earth is *not* a deduction from radiometric dating. Radioactivity was not discovered until the very late 19th century, and radiometric dating techniques were not well developed until the mid 20th century. The old earth is a *belief* that became an *assumption* most scholars accepted in the mid 17th century. In his book on "deep time," pertaining to the old earth concept, Stephen J. Gould (1987, p. 3) writes:

> The acceptance of deep time [an old earth], as a consensus among scholars, spans a period from the mid-seventeenth through the early nineteenth centuries.

Belief in an old earth and universe continued past the early nineteenth century, of course, but Gould was referring to the time before uniformitarianism became accepted. The view that the earth is old came *before* the theory of evolution and uniformitarianism burst onto the scene. In fact, it is very likely that the acceptance of an old age for the earth paved the way for uniformitarianism and the theory of evolution.

On what basis did scholars of the 17th century to the 19th century decide the earth is old? It was mainly because they rejected a global Flood or relegated the Flood to the last of a series of earth catastrophes. Without the Flood, intellectuals concluded that many processes would take millions of years to accomplish. For instance, they looked at large valleys and reasoned that they were eroded by the rivers flowing through them. Of course, by denying the global Flood as the cause of the valley, it would take millions of years to erode valleys by the slow process of river erosion. Early acceptance of an old Earth was bound to bias scientists as more sophisticated means of measuring age were later developed.

HOW GOOD IS RADIOMETRIC DATING?

Radiometric dating is a complicated subject. A full explanation of the assumptions used in this dating method is beyond the scope of this book, but it is important to briefly mention some major problems with the method. I will refer to various new research that the reader can obtain for more information. Radiometric dating is largely based on three main assumptions (Morris, 1994; Vardiman, Snelling, and Chaffin, 2000). First, when one radioactive element (the parent element) decays into another element (the daughter element), the rate of change must remain constant for millions or billions of years. Second, the amount of the parent radioactive element and the daughter element must be *known when the rock formed*. Since radiometric dating methods almost always date igneous rocks, the initial time of formation of the rock would be when the rock solidified from the molten phase, unless there was a heating or metamorphic event since hardening. In this latter case, the "date" may refer to these later events. Third, the rock must remain isolated or in a closed system, away from all leaching and contaminating substances since the time the rock formed. None of the parent or daughter elements must leave or enter into the rock for millions and billions of years. These assumptions cannot be proven, and in many cases they are known to be problematic.

In regard to the first assumption, scientists have not taken into account processes that may have increased the rate of radioactive decay, such as the Creation and the Flood (Humphreys, 2000; Chaffin, 2000, 2001).

With respect to the second assumption, how are scientists to know the amount of the parent radioactive elements and the daughter products when the igneous rock first cooled and solidified? One might expect that at the time the rock cooled, there would be no daughter products of radioactive decay. The heat should have driven away any daughter elements. When evolutionists date, they practically always *assume* some daughter products of radioactive decay at the beginning. For instance, in the potassium argon dating method in which parent radioactive potassium-40 decays to stable daughter argon-40 (a gas), there is almost always some argon-40 assumed in the rock at the beginning. Argon-40, a gas, should have been driven away due to the heat of the liquid rock. It requires further speculation to estimate how much of the daughter product was originally present. The postulated initial parent and daughter elements leaves the radiometric dating systems open to preconceived ideas and makes it possible to pick and choose the dates considered "correct." The third assumption is often violated. One problem is that most of the radioactive elements and their daughter products are *readily removable* by ground water, or by heating episodes. Once this happens there is no reliable way to date the original cooling time of the rock by radiometric dating.

These are the three main assumptions. There are also many other questionable assumptions that must be applied to a particular dating method (Faure, 1986). Besides this there are many difficulties in applying individual dating methods. Even more problematic is the fact that there are a large number of anomalous or "wrong" dates, probably a majority of them, that are rejected because they do not affirm dates already assumed to be correct (Woodmorappe, 1999c; Vardiman, Snelling, and Chaffin, 2000).

Uniformitarian scientists are well aware of the many problems with radiometric dating and often use the problems to reject the many dates that they consider wrong (Woodmorappe, 1999b; Austin, 2000). For instance, they often blame a "wrong" date on contamination, a violation of the third assumption. This rejection is almost always an *after-the-fact* conclusion, since they do not deliberately choose samples that are believed to be contaminated before they date the sample.

Some may object that radiometric dates must be correct because they "agree" with the dates already established by the fossil age of nearby sedimentary rocks. What these people are not unaware of is that a date is considered "wrong" when it does not agree with the already predetermined age or age range of the rock (Snelling, 2000). Of course, radiometric dates would agree with the fossil dates in this situation, *when the fossil date is the criterion for judging a good date from a "wrong" date*. This is an example of circular reasoning.

Occasionally uniformitarian scientists will date the age of a rock into the future or older than the assumed age of the earth (Snelling, 2000). Many recent or ice age lava flows have been dated at millions, and sometimes billions, of years old by various radiometric dating methods (Austin, 1996; Austin and Snelling, 1998; Rugg and Austin, 1998; Snelling, 1994, 1995, 1998, 2000). Since the radioactive date is suppose to date the time the lava cooled, something is wrong. It is likely one or more of the three major assumptions have been violated.

This means that radiometric dating really is *not* the primary dating method. It is the *fossil date*, based on certain index fossils thought to have lived during short periods of a few millions years –this is the primary dating method. The question can legitimately be asked: How old are these index fossils? Their age is based on the assumption of evolution in which organisms supposedly started out simple and became more complex with time. In other words, evolution is the *assumption* repeatedly used in creating and choosing fossil dates as well as other dating methods. Again, this is an example of circular reasoning.

It may come as a shock to many to discover that these highly touted radiometric methods are not as important as the scheme devised using fossils to determine the dates. The fossil dating system was concocted back in the early and mid 1800s. It is easy to verify the use of fossils as the plumb line for good dates from reading the geological literature. I discovered it by accident during my study of the geology of the Pacific Northwest of the United States. Note this statement by the late professor Bates McKee (1972, p. 25) of the University of Washington concerning the geology of the Pacific Northwest:

> One might imagine that direct methods of measuring time would make obsolete all of the previous means of estimating age, but these new "absolute" [radiometric] measurements are used more as a supplement to traditional methods than as a substitute. Geologists put more faith in the principles of superposition [rocks above other rocks are younger] and faunal succession [the change of fossils with time or evolution] than they do in numbers that come out of a machine. If the laboratory results contradict the field evidence [from fossils], the geologist assumes that there is something wrong with the machine date. To put it another way, "good" dates are those that *agree* with the field data [brackets and italics mine, quotes his].

This is quite an admission, but it is common knowledge within the earth sciences that fossils are more important than radiometric dating (Woodmorappe, 1999b).

MANY OTHER DATING METHODS GIVE YOUNG AGES

Besides dating methods that give old dates, there are many dating methods that give much younger dates than those from radioactivity. The reader must be cautioned that even these dating methods are based on assumptions, often the same assumptions used by the uniformitarian geologist. The point is that these dating methods arrive at much younger dates. I will provide two examples.

The amount of salt eroded into the oceans from the continents was once used as a dating method for the age of the oceans. John Joly in 1899 calculated that the amount of salt in the ocean would have accumulated in 80 to 90 million years, which was subsequently considered the age of the earth (Austin and Humphreys, 1990, p. 17). This date was superseded by radiometric dating. The date Joly discovered was explained away as just the time for steady state to be reached in which removal processes balanced the input by rivers. Austin and Humphreys (1990) updated Joly's argument by cataloging all known *and conjectured* input and removal processes. They calculated an upper limit for the age of the ocean of *62 million years* by applying a differential equation containing *minimum* input rates and *maximum* output rates. This figure, therefore, is quite conservative and well below the billions of years assumed by uniformitarian scientists. Furthermore, this number does not even take into account the global Genesis Flood that would have washed copious amounts of salt into the oceans. So the real age of the ocean is very much less than 62 million years.

The decay of short period comets indicates that the solar system is quite young (Faulkner, 1998). Comets are considered "dirty snowballs." Every time a comet goes around the sun in a short-period orbit of less than 200 years, it loses matter. Since comets are small objects to begin with, about 6 miles (10 kilometers) in diameter, it would not take too long since the origin of the solar system before all short-period comets would have disappeared. The amount of time depends on several variables but it can be calculated in thousands of years. The fact that there are around 100 short-period comets left indicates that the solar system is thousands of years and not billions of years old. The Oort cloud of comets supposedly located about 100,000 astronomical units away from the sun is supposed to supply the solar system with short-period comets over billions of years. An astronomical unit is the distance between the sun and earth. However, the Oort cloud has not been observed!

Because of many problems with the Oort cloud hypothesis, only postulated as an explanation for the short age of the solar system, astronomers now favor the origin of short-period comets from what is called the Kuiper belt. This belt starts from just beyond the orbit of Neptune to well beyond the orbit of Pluto. Unfortunately, the objects observed by astronomers in the Kuiper belt, so far, are much larger than comets. Even Pluto is considered by some astronomers to be a Kuiper belt object. So, it is speculative to assume that comet-sized objects, too small to be seen in telescopes, exist in the Kuiper belt. Besides it is problematic, if comets did exist in the Kuiper belt, for them to end up in a highly elliptical orbit that extends from around the sun to near the orbit of Jupiter. Thus, there does not appear to be any mechanism to generate short-period comets, while all we observe is the number of comets *diminishing* with time. We observe them dimming, being ejected from the solar system, colliding with Jupiter, and disintegrating. The straightforward evidence of short period comets indicates that the solar system is very young!

TEPHROCHRONOLOGY, CARBON-14, AND THE DATE OF THE LAKE MISSOULA FLOOD

It was the ice age dating method called tephrochronology, bolstered by the carbon-14 dating method that was used to date the Lake Missoula flood (Mullineaux *et al.*, 1978; Mullineaux, 1986; Moody, 1987, pp. 78-79). Tephra is material ejected from a volcano into the air. Tephrochronology uses various chemical and other signatures of a tephra to arrive at a date. However, the date must be derived by *other* dating methods first. In the case of the Lake Missoula flood, it was the carbon-14 dating method that was originally used to date the tephra. Carbon-14 is used to date organic remains, but its range is only a maximum of 50,000 years within the uniformitarian dating scheme.

To obtain a date for the Lake Missoula flood, scientists chemically analyzed the volcanic ash in the slackwater rhythmites of south central Washington. Then they deduced the

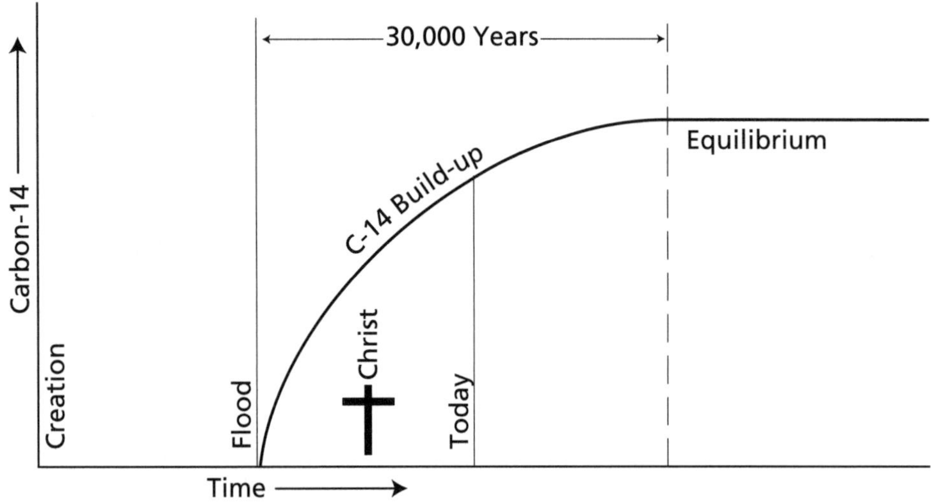

Figure 8.1 Schematic of the carbon-14 dating system using the Flood model and the disequilibrium of the system (from Morris, 1994, p. 64).

particular mountain from which the ash originated, and the particular eruption from that mountain. The ash in the slackwater rhythmites was correlated with "set S" ash from Mount St. Helens. Once this determination was made, organic material associated with the set S ash near the mountain was dated by carbon-14. This date turned out to be about 13,000 years (Mullineaux *et al.*, 1978). How solid is this date from tephrochronology and carbon-14, which is older than the time scale allowed by Scripture?

First, there are many difficulties in the tephrochronology dating system, and the Cascade volcanoes are no exception. Begét, Keskinen, and Severin (1997, p. 140) admit: "Much remains to be learned about the age and distribution of tephras produced by major Pleistocene eruptions of the Cascade volcanoes." Moody (1987, p. 94) acknowledges many problems with the use of tephrochronology in the Channeled Scabland:

> The above discussion on the identification of tephra layers within the study area demonstrates that chemical and petrographic data can not and must not be used alone for the identification of volcanic ashes. The stratigraphic and sedimentologic date must also be used to distinguish between many similar ashes [emphasis hers].

She not only admits to the similarity of many ashes, but also that *previous beliefs* about the timing of the sediments should dictate the approximate dates. Again circular reasoning enters into the dating scheme. Westgate and Naeser (1985) corroborate Moody's assessment in admitting that making distinctions between similar ashes is sometimes difficult and a *multivariable approach* is needed.

There have been about 100 eruptions of Mount St Helens (Mullineaux, 1986), and the older "set M" ash is particularly similar to the set S ash (Mullineaux *et al.*, 1978; Begét, Keskinen, and Severin, 1997, p. 142). Set S was chosen over the set M as the ash near the top of the Lake Missoula flood rhythmites. One wonders how objective was such a procedure. Regardless, it was the carbon-14 method that is really responsible for the date of 13,000 years ago.

Second, the carbon-14 method is touted as being quite accurate, but it is not. It is based on the usual three main assumptions of radiometric dating, plus a few more. In my 30 years of studying the ice age, I have come across hundreds of carbon-14 dates, at first believed to be correct, that were later rejected because of "contamination," lack of agreement with other information, or some other reason. So, the method could not really be as good as some scientists claim.

Other carbon-14 dates of the "last" Lake Missoula flood illustrate the problems with the method. The last Lake Missoula flood was at first dated at 33,000 years ago by the carbon-14 method (Fryxell, 1962; Mullineaux *et al.*, 1978, p. 171). This date was later rejected because the standard glacial chronology at the time suggested a date of about 20,000 years (Bretz, 1969; Mullineaux *et al.*, 1978, p. 171). But many scientists thought the Lake Missoula flood was even younger than 20,000 years (Mullineaux *et al.*, 1978), based again on assumed glacial chronology. This is the ever-present subjective element that good carbon-14 dates are those that *agree* with preconceived opinion. This is a good example showing how preconceived ideas of past events enter into the choosing of "dates." It seems the carbon-14 dates have followed the latest thinking on the date of the "last" ice age. Circular reasoning abounds.

To add more confusion, Begét, Keskinen, and Severin (1997) further claim that the 13,000-year date for the set S ash is now *too young*, and they would like to move the date back to around 15,000 to 16,000 years ago. This request is based on the glacial chronology from the Puget Sound area.

Still the carbon-14 dating method produces dates much greater than 6,000 years –the Scriptural time scale. That is because they assume the carbon-14 method is in equilibrium, but it is not. For equilibrium, the amount of carbon-14 being formed in the atmosphere by cosmic rays must balance the carbon-14 being changed back to nitrogen-14 by radioactive decay. Apparently, there is a lack of balance by about 20%, which likely means that the Earth is less than 30,000 years old (Morris, 1994, pp. 64-67). If carbon-14 production was recently initiated, it would take 30,000 years for the decay to balance the production. This imbalance implies that the earth is less than 30,000 years old!

In the carbon-14 dating method, it is the carbon-14/carbon-12 ratio that is actually used to provide a date.

Carbon-12 is the normal, most abundant isotope of carbon that is associated with living organisms. However at the time of the Genesis Flood, this ratio would be greatly disturbed and as a result would not fit into the assumptions made by geochronologists. The organic disturbance was caused by the death and burial of a huge amount of organic matter that contained carbon-12, and then the subsequent great increase in carbon-12 after the Flood when the world was rapidly repopulated with plants and animals. If we use the assumption of the Flood, a different carbon-14/carbon-12 ratio than today, and the disequilibrium of the carbon-14 system, we can telescope "reliable" carbon-14 dates within the time scale of the Flood (Morris, 1994, p. 64-67) (Figure 8.1).

ARE OTHER ICE AGE DATING METHODS RELIABLE?

There are many newer dating methods applied to the ice age. One would think that these newer methods would provide a better date than the carbon-14 method for the Lake Missoula flood and other ice age events. In the opinion of some, *all* of these ice age dating methods are inferior to carbon-14. Karrow and others (1997, p. 99) lament:

> The problems of determining the age of the fossils at this site are fundamental and common in Quaternary [ice age] stratigraphy, particularly for sites older than 40 kyr [40 thousand years]. Dating methods other than radiocarbon are *not yet reliable nor widely accepted and used* [emphasis and brackets mine].

As we have seen, the carbon-14 method is problematic.

Furthermore, it is a common practice to calibrate one dating method with another, which supposedly makes many of these dating techniques "consistent" with each other, but it takes away from the independence of each dating method.

Regardless, carbon-14 then is accepted as the best ice age dating method; the others are considered inferior or simply have been adjusted to produce similar dates to carbon-14. Since carbon-14 is not that reliable and can be adjusted by using different assumptions to within the time of the Flood, ice age dating methods are no threat to the Scriptural time scale.

OBJECTIVE EVIDENCE THAT RADIOMETRIC AND FOSSIL DATING METHODS ARE WRONG

Modern erosion rates and certain ephemeral landforms provide objective evidence that radiometric and fossils dates cannot be correct (Oard, 2000b).

Geomorphologists, who study the shape of the land surface, up until recent times believed that most landforms were no older than Pleistocene or, at most, late Tertiary within the uniformitarian geological time scale. These are the last epochs of geological time and represent only several million years. This belief is because current weathering and denudation rates of the continents are relatively fast (Roth, 1998, pp. 263-266; Schumm, 1963). Consequently, no landform should exist for more than several million years.

However, over the years, geologists have dated some landforms tens of millions to occasionally *over one hundred million years old* based on various radiometric and fossil dating techniques. These old landforms are mostly planation or erosion surfaces (Figures 8.2) and sometimes river valleys (Oard, 1996b,c, 1997b, 1998a). One erosion surface on the Kimberly Plateau, Western Australia, is believed to have been planed 600 million years ago with *little* erosion since and with apparently no sediments to cover and protect the surface for all this time (Ollier, Gaunt, and Jurkowski, 1988)! Another example is the flat to undulating plateau of western Arnhem Land, Queensland, Australia. It is dated at over 100 million years old based on late Jurassic and Cretaceous fossils. The fossils are found in sedimentary rocks that lie in shallow valleys that have been cut after the erosion surface was planed (Nott and Roberts, 1996). These fossils give a minimum age to the erosion surface. Based on K-Ar dating of basalt lava that had flowed into the ancestral Shoalhaven River Gorge of southeast Australia, uniformitarian geologists were surprised to find that the walls of the gorge retreated

Figure 8.2 Flaxville cobble and boulder capped planation surface near Scobey, northeast Montana. Cobbles and boulders have been transported from the Rocky Mountains. This surface is the second from the top of four general levels of planation in the region.

only 33 feet (10 meters) in 30 million years (Nott, Young, and McDougall, 1996)! The river is considered ancient but its width has changed little over many millions of years, according to their dating technique. How can the walls of a gorge eroded so slowly? Do these exceedingly slow rates of erosion make sense?

Although many geomorphologists remain unconvinced of the great antiquity of landforms, this "extremely unlikely" concept has been supposedly vindicated, according to Australian geomorphologist, C.R. Twidale (1998). Not only are some erosion surfaces in Australia and Africa much older than 100 millions years by uniformitarian reckoning, but old planation or erosion surfaces (palaeosurfaces) are a *worldwide* occurrence. Twidale (1998, p. 657) states:

> Yet for the past half century or more palaeosurfaces have been recognized and compelling evidence adduced pointing to their great antiquity, not only in Australia and Africa but also, and in lesser measure, in the Americas and Europe.

The task is now "...to account for the seemingly impossible..." (Twidale, 1998, p. 662). He rejects William Morris Davis's "cycle of erosion" and other such cyclic schemes and leans towards Crickmay's "hypothesis of unequal activity"(Crickmay, 1974; 1975), which Twidale admits only *diminishes* the problem without solving it (Twidale, 1998, p. 663).

Davis's "cycle of erosion" was immensely popular during the first half of the 20th century, but it is mostly rejected today because it is hypothetical. It is *not* based on observable processes, and there are no current examples of a "peneplain," the final product of erosion, forming today at sea level (the ultimate base level). The cyclical schemes of Lester King and Walther Penck have fared no better (Summerfield, 1991, pp. 457-480). These cyclical hypotheses were attempts to account for the many planation surfaces observed over the earth.

Crickmay essentially believes that rivers account for most of the erosion of continents and the erosional activity is unequal. He is correct, but it is not particularly enlightening. Crickmay's hypothesis is supposed to account for the survival of high level planation surfaces formed by water. However, today these planation surfaces are observed to be weathering and eroding. So, although the rivers can erode faster (unequal activity), his hypothesis still does not account for old landforms originally formed by water but barely touched by erosion over "tens of millions of years" ever since. Even a slower erosion rate of these planation surfaces should soon destroy them within uniformitarian geological time.

Crickmay created his "hypothesis of unequal activity" because he recognized the contradiction between the dates of planation surfaces and current weathering rates and realized that current hypotheses failed to account for old surfaces. He states the problem this way:

> Again, one finds all over the world, even high above and far distant from existing waterways, smooth-surfaced and level ground - including everything from small terraces to broad, flat plains –much of it still bearing intact a carpet of stream alluvium. Such lands were carved and carpeted, evidently, by running water, even though they are now in places where no stream could possibly run. What is remarkable about them is the perfection with which they have outlasted the attack of 'denudation' for all the time that has passed since they lay at stream level (Crickmay, 1974, p. 173). The cobble and boulder capped planation surfaces are worldwide and often at high elevations. The cobbles and boulders are often rounded indicating that a process involving water cut the planation surface and left behind the rocks.

It really is against common sense that these planation surfaces can be tens of millions to over a hundred million years old, as admitted by Twidale (1998, p. 664):

> If some facets of the contemporary landscape are indeed as old as is suggested by the field evidence they not only constitute a denial of commonsense and everyday observations but they also carry considerable implications for general theory.

Twidale and others continue to fish around for mechanisms to preserve these "old" surfaces. A resistant rock cap such as a hard sandstone or a duricrust is one possibility. Resistant rocks would indeed slow erosion, but likely not enough to last as long as postulated. The fact that planation surfaces have been planed evenly on both tilted hard and soft sedimentary rocks indicates that more than structure is involved (see Figure 7.12).

Especially contradictory to their "old" age is that some truncated surfaces still exist that were cut on relatively *soft, easily erodible* rocks (Twidale, 1998, p. 663; Crickmay, 1974, pp. 207,209). One would expect soft rocks to easily form a drainage network that would soon destroy the flatness. These surfaces must be young.

Some geologists appeal to a dry climate as a preserving mechanism, but during supposed geological time, planation surfaces are expected to have passed through several climatic regimes. Australia supposedly has been slowly drifting northward from the high precipitation mid and high latitudes during the past 100 million years of geological time. Although much of southern and central Australia has a dry climate today, these areas would have been much wetter during the Tertiary. Besides, erosion is not suspended in a dry climate. Semi-arid climates are well known for occasional heavy rain from thunderstorms that cause great erosion. Summerfield (1991, p. 396) lists average denudation rates for various climates and relief, based on both the solid and dissolved load of major rivers today. A landscape in a dry climate with low relief denudes at roughly 0.2-1.4 inches/1000 years (5-35 mm/1000 years). This is significantly fast for the supposed age of these landforms. Flat or nearly flat planation surfaces would not be expected to last long.

Twidale seems to be desperate for explanations when he appeals to *glacial protection* in areas once covered by ice sheets (Twidale, 1998, p. 663). Geologists who study the ice age now realize that ice caused little erosion, except in local areas, such as fjords and valleys (Lidmar-Bergström, Olsson, and Olvmo, 1997). Since some erosion surfaces *survived* the ice age, Twidale suggests that a thin veneer of debris helped to preserve these erosion surfaces.

The survival of planation surfaces all over the earth is objective evidence that the dating methods responsible for the "old ages" are highly exaggerated. This justifies the search for other interpretations for radiometric dating methods, as exemplified by the massive research effort by the creationist group called RATE (Radioisotopes and the Age of the Earth), a joint effort by the *Institute for Creation Research* and the *Creation Research Society*.

As discussed in Chapter 7, planation surfaces speak better of a mechanism that occurred in the past, but which is no longer in operation today. It has to be a worldwide mechanism, since planation surfaces are seen all over the earth, even on Antarctica. The mechanism was large scale, able to quickly shear hard and soft rocks evenly, and then erode the entire rock mass further so that planation surfaces are mostly left as remnants. It had to be a *watery* catastrophe based on the rounded rocks capping many planation surfaces. Furthermore, it was the last major event to shape the surface of the land before erosion from the present climate began to slowly dissect them. Finally, it occurred fairly recently. The mostly likely mechanism to explain our present geomorphology is the Recessive Stage of the Genesis Flood as the waters drained off the land (Walker, 1994).

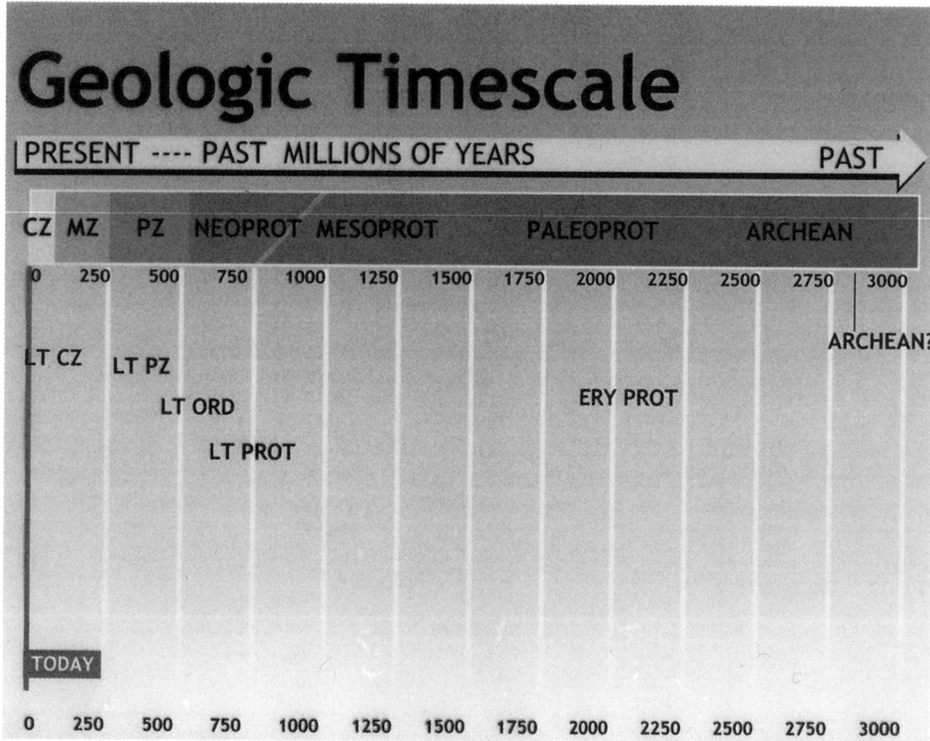

Figure 8.3 Schematic of the five main ice age periods in uniformitarian earth history. The Archean "ice age" is still speculative. The Pleistocene ice age is referred to as the late Cenozoic (LT CZ) because Antarctic supposedly developed in the mid and late Cenozoic, while the other ice sheets developed in the late Pliocene of the geological time scale (redrawn from John Crowell, 1999 by Dan Lethia).

Figure 8.4 "Tillite" from the late Precambrian "ice age" near Pocatello, Idaho. Notice the random stones embedded in a finer-grained matrix.

DOES AN ICE AGE TAKE MILLIONS OF YEARS?

Besides the old ages derived from radiometric dating methods, there are many earth processes that are claimed to take many millions of years –much too long for the short Scriptural time scale. I will choose one process that I have analyzed in depth to show that the assumed time for such a process is based on how one views the earth. This process is

Figure 8.5 Striated pavement from Permian "ice age" Hallett Cover, just south of Adelaide, Australia (photo courtesy of Andrew Snelling).

Figure 8.6 A volcanic debrite, a lahar, in the Ellensburg Formation, 10 miles (16 kilometers) west of Ellensburg, Washington. The lahar flowed out of the Cascade Mountains of Western Washington.

the ice age or ice ages. In the uniformitarian scheme, ice ages not only repeat at regular intervals over the past 2.5 million years, but also occur at various times during geological time (Figure 8.3). The former are referred to as the Pleistocene or Quaternary ice age (although they also stretch into the late Pliocene), while the latter are considered pre-Pleistocene or pre-Cenozoic ice ages, which extend back past 2 billion years ago in evolutionary time. The Pleistocene ice age represents 30 or more successive ice ages (Kennett, 1982, p. 747). Each ice age supposedly repeats at 100,000-year intervals for the past million years and at about 40,000-year intervals before that. In these repeating ice ages during the past million years of uniformitarian time, 90,000 years represents a glacial period, while 10,000 years are times of interglacial moderation. Uniformitarian scientists think we now live in a rapidly fading interglacial period called the Holocene. A pre-Pleistocene ice age is said to often last over 100 million years and even extend into the tropics. Did all these ice ages actually occur? Did they last as long as the scientists say? If the above estimates are even close, the Scriptural time scale is very wrong.

I will briefly discuss another side to this story, based on a different view of earth history. Since I have written in-depth studies for both the Pleistocene ice age (Oard, 1990) and the pre-Pleistocene "ice ages" (Oard, 1997a), I will only summarize those works here. The readers who want more information can consult the above two studies. I will deal with the pre-Pleistocene ice ages first.

PRE-PLEISTOCENE "ICE AGES"

The rocks that suggest pre-Pleistocene ice ages are found within thick masses of sedimentary rocks. The "ice age" layer is composed of stones inside a fine-grained matrix (Figure 8.4). When such a layer is derived from a glacier it is called *till* (see Figure 8.12). When the layer is consolidated, it is called a tillite. Sometimes seeming ice age features, such as striated pavements (scratches on bedrock surfaces) (Figure 8.5) and striated rocks (scratched cobbles or boulders) are also found. There is little doubt that these rocks are part of the great accumulation of sedimentary rocks laid down by the Flood. It is obvious that an ice age could not occur within a one year Flood. Anti-creationist, Arthur Strahler (1987, p. 263) offers the following challenge:

> The Carboniferous tillites cannot be accepted by creationists as being of glacial origin for the obvious reason that the tillite formations are both overlain and underlain by fossiliferous strata, which are deposits of the Flood. During that great inundation, which lasted

the better part of one year, there could have been no land ice formed by accumulation of snow.

Strahler is correct that there could be no ice age during the one year Flood.

There is strong evidence, however, that they have misinterpreted the rocks, as even a few uniformitarian geologists have pointed out (Schermerhorn, 1974). A mechanism not considered until recently can account for these till-like deposits. This mechanism is mass movement, or in other words *gigantic landslides*. Two main types of landslides are debris flows and turbidity currents. A debris flow is a moving mass of rock fragments of all sizes within a finer-grained matrix. A turbidity current is similar, but generally with fewer rocks and more water within the moving mass. The underwater flow of sediment is supported by fluid turbulence. When debris flows and turbidity currents stop moving they are called debrites and turbidites, respectively.

Debrites and turbidites can mimic real ice age features. Debrites mimic ice-age deposits called till that contains rocks of various sizes mixed within a fine-grained matrix (Figure 8.6). Striations result when rock scrapes against rock or bedrock during a mass flow. I have examined a debris flow found on top of the Gravelly Mountains of Southwest Montana. The flow resulted in isolated striated rocks and a well-defined striated pavement (Figure 8.7) that has since been broken up by subsequent slumping. These features were not made by glaciation, although at one time the deposit was interpreted as an ancient glacial deposit. Fine-layered turbidites can look just like another glacial feature, silt/clay rhythmites with rocks dropped from icebergs in a floating lake.

The Genesis Flood with its regional-scale rapid sedimentation can explain these pre-Pleistocene "ice age" deposits. Some areas of rapid sedimentation would likely become unstable during deposition, and because of tectonics or giant earthquakes during the Flood, the mass would move down even a gentle slope. These mass movements would be on a large scale, which can account for the huge size of some of these supposed ice age rocks. Since the Genesis Flood can account for these unique rocks in this way, the many millions of years suggested by pre-Pleistocene "ice ages" disappear. This goes to show that *a different view of earth history with the addition of other variable or a new mechanism can change the entire chronological picture*. What some scientists view as a hopeless contradiction to the time scale of Scripture

Figure 8.7 Debrite with a striated boulder sitting on top of a striated pavement on top of the Gravelly Range Mountains of Southwest Montana. Striated boulders and pavements were once thought to be uniquely diagnostic of glaciation.

Figure 8.8 Striated pavement in the Sun River Canyon of the Rocky Mountains west of Great Falls, Montana.

turns into support for a short time scale during the Flood. The Flood often provides a solution to these apparent time contradictions. This alternative theory to pre-Pleistocene "ice ages" is presented in the study, *Ancient Ice Ages or Gigantic Submarine Landslides?*

THE PLEISTOCENE ICE AGE

The Pleistocene ice age is characterized by unconsolidated debris that is composed of many sized stones and mixed in a fine-grained matrix. This debris called glacial till, lies on the surface of many areas at the mid and high latitudes. Landslides cannot account for this debris over such a huge distance of low relief. Striated pavements upon which the ice sheets slide are relatively common (Figure 8.8), as well as striated boulders in moraines (Figure 8.9). Erratic boulders often transported in icebergs on lakes or meltwater-gorged rivers, lie within and just south of the area once occupied by the continental ice sheets. The Belleview erratic in the Willamette Valley (see Figure 2.11) is an example of a large boulder that was transported well south of the ice sheet by an iceberg during the Lake Missoula flood. Many mountain areas of the mid latitudes and tropics possess glacial features at much lower elevations than current permanent snow lines. Distinctive horseshoe-shaped moraines are found at the entrances to mountain valleys all over the western United States (Figures 8.10 to 8.12). Such geometric shapes could not form during the Flood, so this ice age was real and *followed* the Flood.

Interestingly, the deposits and other features of the Pleistocene ice age commonly appear to be fresh with little sign of weathering, suggesting that the ice age happened recently. Moraines, drumlins, eskers and other glacial features for the most part are still sharp and little eroded. Notice the sharp lateral moraine surrounding Wallowa Lake in Figures 8.10 and 8.11. Wright (1911, p. 569) also recognized that striations around Hudson Bay have experienced little weathering:

> On Portland promontory, on the east coast of Hudson's [sic] Bay, in latitude 58^0, and southward, the high, rocky hills are completely glaciated and bare. The striae are as fresh looking as if the ice had left them only yesterday. When the sun bursts upon these hills after they have been wet by the rain, they glitter and shine like the tinned roofs of the city of Montreal.

Since the ice age was the last major event of earth history, it should be a showcase for uniformitarian geology, in other words one would expect that the ice

Figure 8.9 Striated boulder from a terminal moraine on the high plains west of the Sun River Canyon in the Rocky Mountains, Montana.

Figure 8.10 Horseshoe-shaped lateral and terminal moraine extending out onto small plain with overdeepened Wallowa Lake filling the depression between moraines, northern Wallowa Mountains, northeast Oregon. Notice the moraines are sharp crested, indicating their youth.

age would be relatively easy to explain by mainstream geologists. When you read any book on ice age geology, it is easy to get the impression that uniformitarian scientists can indeed explain the ice age. In reality, the origin of the Pleistocene ice age remains a *major mystery* of earth science in spite of over 160 years of research (Oard, 1990). Daniel Pendick (1996, p. 22) writes: "If they hadn't actually happened, the ice ages would sound like science fiction." This is because many atypical conditions have to be in place at the same time for an ice age to develop. Yet, evidence for an ice age is widespread. The cause of the ice age was declared to be one of the 18 great mysteries of science in the August 18-25, 1997, *U.S. News and World Report* (Watson, 1997).

Figure 8.11 Eastern lateral moraine of Figure 8.10. Moraine is about 600 feet (183 meters) high.

Many hypotheses or theories have been proposed to explain the ice age. As of 1968, 60 theories had been considered (Eriksson, 1968, p. 68). However, all of them have serious problems, indicating the difficulty of the problem. Ice age researcher, J.K. Charlesworth (1957, p. 1532), proclaimed:

> Pleistocene [ice age] phenomena have produced an absolute riot of theories ranging 'from the remotely possible to the mutually contradictory and the palpably inadequate.'

Figure 8.12 Close up view of the till within the eastern lateral moraine of Figure 8.11. Notice the rocks of various sizes in the fine-grained matrix.

This was back in 1957, but the situation has not changed. David Alt (2001, p. 180) states: "Although theories abound, no one really knows what causes ice ages."

The conditions necessary to form an ice age are major hurtles: 1) much cooler summers (winters in most areas are cool enough); 2) huge amounts of snowfall; and 3) the persistence of the mechanism for hundreds of years or more. The needed cooling can be grasped from a computer simulation of snow cover for one summer in eastern Canada (Williams, 1979). Williams attempted to demonstrate how substantial summer cooling is necessary for an ice age to develop in northeastern Canada. His simulation resulted in the winter snow of at least one inch remaining one summer in northeast Canada. To accomplish this, he decreased the average summer temperature by increments of two degrees with double the average winter snowfall. Figure 8.13 presents his results with a summer temperature decrease of 18^0 and 20^0F (10^0 and 12^0C).

Williams' simulation was just for northeast Canada, but the ice sheet developed mostly *in place*, as far south as the northern United States. The periphery of this ice is expected to be dynamic with many southward surges (Oard, 1990). For an ice sheet to develop in the northern United States, the

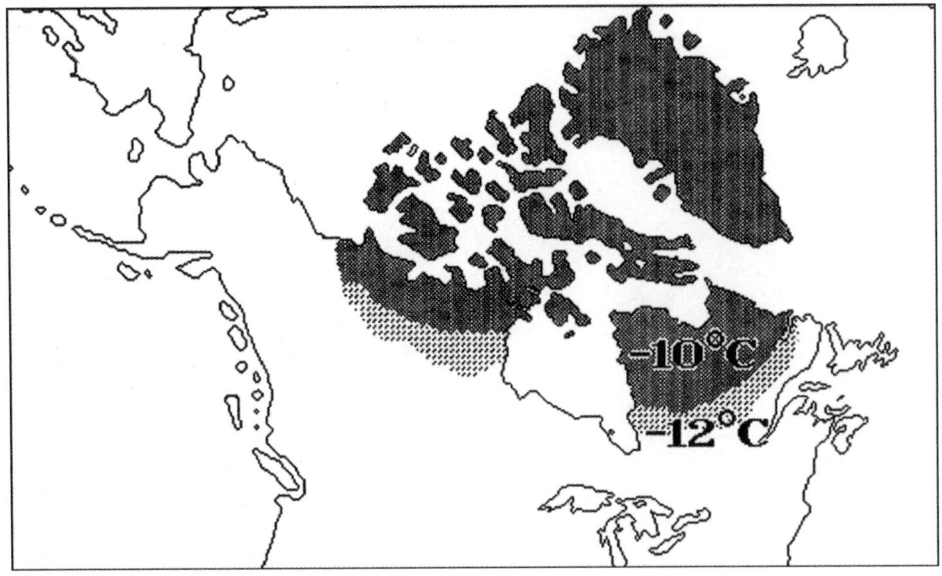

Figure 8.13 Williams' simulation of the amount of temperature change and area for one inch of winter snow to remain after summer melting in eastern Canada (Taken from Williams, 1979).

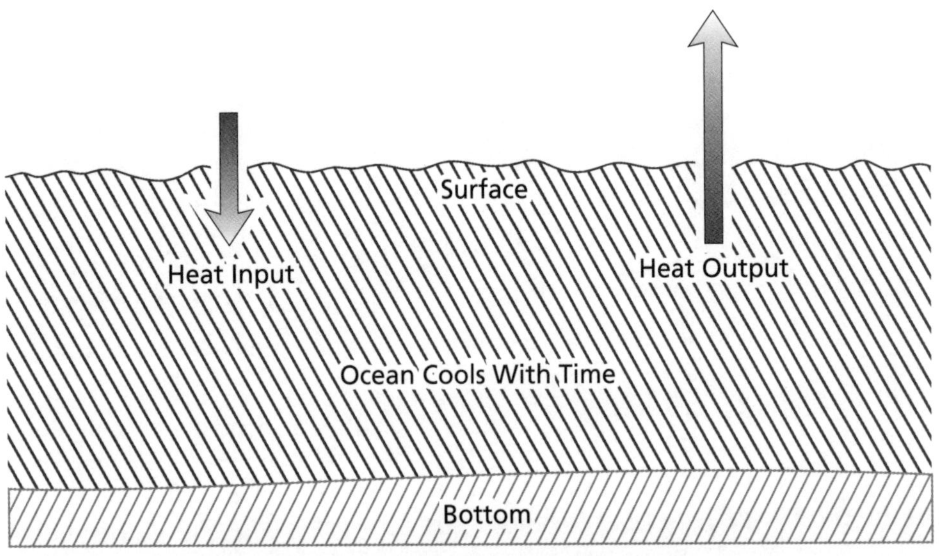

Figure 8.14 Balance of heat inputs and outputs in the ocean determines the cooling time of the ocean (drawn by Dan Lethia).

average summer temperature of a city such as Minneapolis would have to drop to at least 32°F (0°C) from an average around 70°F (21°C). But this is not sufficient cooling. The summer temperature must cool much below the freezing mark since the main cause of melting snow and ice is summer sunshine (Paterson, 1981, p. 313), which lasts long at Minneapolis. Pickard (1984) notes that net summer melting *begins* in Antarctica when the temperature warms up to 14°F (-10°C). Since summer sunshine in Minneapolis is more intense than in Antarctica, the average summer temperature would most likely need to drop below 14°F (-10°C), but a conservative threshold would be down to 20°F (-7°C). This represents a conservative 50°F (28°C) drop from today's average summer temperature! The fact is our present climate is incapable of producing such a huge summer temperature anomaly.

And even if a summer cooling mechanism conceivably could be found, abundant snowfall is also needed –the second criterion. This causes another major problem in that the cooler the air, the less it is able to hold moisture –the opposite of what is required. Uniformitarian enthusiasts are left with a serious problem: What could cause such a drastic climate change?

Williams' experiment was just for one summer. But, this climatic anomaly had to *persist for hundreds of summers* for an ice sheet to develop. So you can see why the ice age is really a major mystery of science.

Not only is its cause shrouded in mystery, but also there are many subsidiary questions associated with the ice age. Why was there consistently a driftless area (an area that was not glaciated) in southwest Wisconsin? This becomes harder to explain when we consider that scientists believe 30 or more ice ages have developed and melted during the past 2.5 million years. Surely, one of these ice ages should have spread over southwest Wisconsin. Another question is: Why and how did millions of woolly mammoths and other large mammals live in Siberia and Alaska during the ice age? Why are some of their carcasses found frozen in the permafrost? Why were normally dry areas today, such as the southwest valleys of the United States and the Sahara Desert, wet during the ice age? A good example of one of these lakes is Lake Bonneville (see Figures 6.1 and 6.2) that was 8 to 12 times (depending upon the modern expansion and contraction of the lake caused by annual precipitation) and 800 feet (245 meters) deeper than the Great Salt Lake (Smith and Street-Perrott, 1983; O'Connor, 1993, p. 5). Also, why do ice age animals and plants commonly display a mixture from warm and cold climates in the same area (Graham and Lundelius, 1984, p. 224)? Why did so many large mammals go extinct or disappear from whole continents at the end of the ice age (Martin and Klein, 1984; Ward, 1997)?

Although uniformitarian scientists have not been able to discover the cause of the ice age, nor have they found adequate solutions to its subsidiary mysteries, creationists have a viable mechanism. The Flood involved unprecedented, widespread volcanic and tectonic activity. After the continents and mountains rose out of the waters, a shroud of volcanic dust and aerosols remained, obscuring part of the sun. This would cause the land to cool dramatically. The dust and aerosols would replenish themselves for hundreds of years following the Flood due to continued volcanism as the earth moved toward equilibrium. There is a great amount of evidence for extensive volcanism within ice age sediments (Charlesworth, 1957, p. 601).

Moisture for snow would come from an ocean that had been warmed by volcanism, lava flows, the friction of tectonics, and possibly water from the crust during the eruption of the "fountains of the great deep" in the Flood. The warm water would be well mixed during the turbulence of the Flood. The warm water extended from pole to pole and from the top of the ocean to the bottom. A well-known principle that affects the formation of clouds and precipitation is the warmer the water, the greater the evaporation. Such a warm ocean would evaporate huge amounts of moisture at mid and high latitudes, only to be picked up by storms and dumped as snow on the nearby continents. This powerful evaporation would continue for hundreds of years until the oceans cooled.

The evaporation would also cause the ocean water to cool with time. I estimated heat inputs and outputs and was able to arrive at a ballpark figure for the length of time it would take for the ocean to cool and an ice age to form (Figure 8.14). Since the balance equations for the ocean and atmosphere would be speculative and applied to a cooler, wetter ice age climate, I used minimum and maximum values for the variables. From this I calculated a maximum and minimum time required to reach glacial maximum. Based on the cooling time for the ocean, the amount of time necessary to reach glacial maximum is a minimum of 174 years and a maximum of 1765 years. Using values in the mid range of the variables, I calculated about 500 years to reach glacial maximum (Oard, 1990).

Regardless of which values are used for the variables in the heat balance equations, the ice sheets would have developed in a very short time compared to uniformitarian estimates. Based on the estimated proportion of moisture available to fall on the ice sheets, I obtained a minimum depth of 1700 feet (520 meters) and a maximum depth of 2970 feet (905 meters) for the Northern Hemisphere. Using variables for the mid range, I estimated an average depth of 2300 feet (700 meters). For Antarctica the best estimate was 3900 feet (1190 meters). The ice sheet would be thicker or thinner than average in various areas, depending upon the distance from the main moisture source, which is the warm ocean, and the distance from the main storm tracks. All of these factors would favor a thick ice sheet in the Pacific Northwest, as well as in southeast Canada. It would predictably be thicker along the southern periphery of the Cordilleran Ice Sheet than at the center of interior British Columbia. So an ice thickness between 2500 and 5000 feet (760 and 1525 meters) in the valleys of northern Idaho, trapping glacial Lake Missoula, is reasonable according to a post-Flood rapid ice age model.

These thickness values are less than uniformitarian estimates. However, the thicknesses of past ice sheets are really only a guess, despite the confidence of some glacial geologists, who generally have *assumed* that past ice sheets would have been generally as thick as Antarctica. Bloom (1971, p. 367) writes:

> Unfortunately, few facts about its thickness are known...In the absence of direct measurements about the thickness of the Laurentide ice sheet [in central and

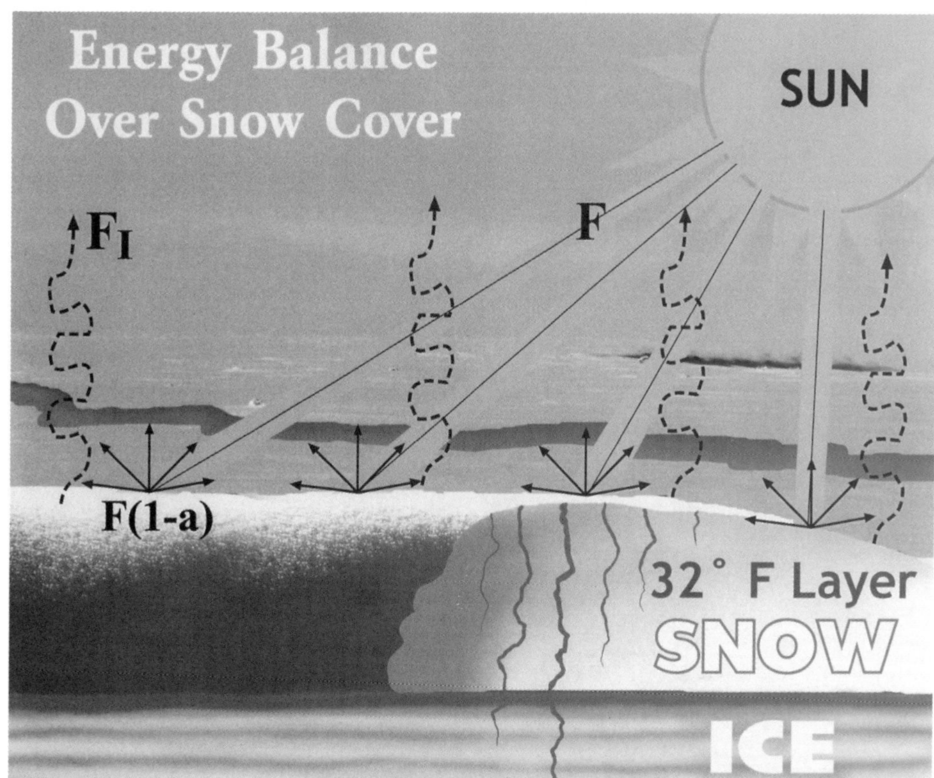

Figure 8.15 Energy balance over a snow and ice cover. F stands for solar radiation, F_I represents the net infrared radiation loss, F(1-a) is the solar radiation absorbed by the ice sheet (the rest reflected), and "a" is the albedo. Note that some water flows like a river on top of the ice, while some percolates down through the 32°F layer, below which is ice (drawn by Dan Lethia).

eastern Canada], we must turn to analogy [Antarctica] and theory.

It is now confirmed that the Laurentide Ice Sheet was thinner than earlier expected along most of its southern periphery (Mathews, 1974; Clayton *et al.*, 1985; Beget, 1986b) and its northwest periphery (Beget, 1987). Furthermore, the interior of the ice sheet was likely multidomed and thinner (Shilts, Cunningham, and Kaszycki, 1979; Shilts, 1980). If John Shaw and colleagues are correct about gigantic subglacial floods, there would be a lake near the center of the ice sheet at peak glaciation and early deglaciation. A thinner ice sheet lends support to the post-Flood, rapid ice age with its significantly thinner ice sheet than postulated by uniformitarian scientists.

Using the energy balance equation over a snow or ice cover (Figure 8.15), I discovered the ice sheets in the Northern Hemisphere would melt in about 100 years near the periphery and about 200 years in the interior of the ice sheets (Oard, 1990, pp. 109-133). This melt rate compares favorably with the current melt rates in the ablation areas of Icelandic, Alaskan, and Norwegian glaciers (Sugden and John, 1976, p. 39). So the ice sheets melted catastrophically. There is even room to account for Shaw and colleagues subglacial flood hypothesis (Shaw, 1996, 2002) in the creationist model. Rapid melting and possibly even a thinner ice sheet towards the center of the Laurentide or Cordilleran Ice Sheets could result in the ponding of large lakes well behind the periphery. After a huge volume of water collected for many years, the lakes could have burst from underneath the ice sheet periphery as subglacial floods.

Figure 8.16 Worldwide distribution of dinosaur footprint discoveries. About 1,500 locations have been known to yield dinosaur tracks.

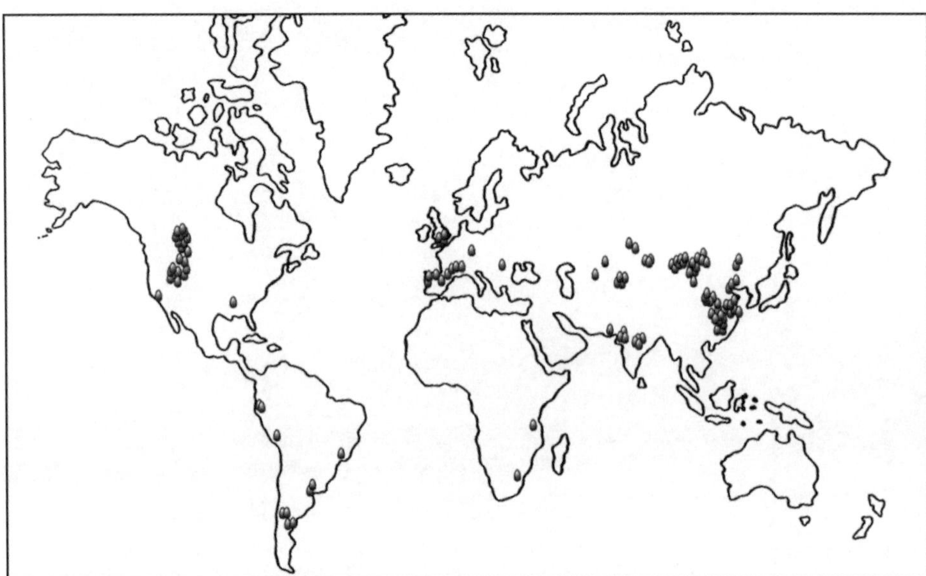

Figure 8.17 Worldwide distribution of 199 sites where dinosaur eggs have been found. Major deposits are few.

Within the Flood model there could only be one ice age. Most of the evidence for multiple ice ages during the Pleistocene is a result of multiple assumptions (Oard, 1990, pp. 135-166). One of the main assumptions is acceptance of the astronomical theory of the ice ages, which asserts that the Earth bounced back and forth between glacial and interglacial periods due to variations in the Earth's orbital geometry. There are a number of serious problems with this theory (Oard, 1990, pp. 15-18). It only accounts for the oscillations, not the beginning of the ice age cycle itself. Especially problematic is that the change in high latitude sunshine, the most important variable, is too small to cause glacial/interglacial oscillations (Paltridge and Platt, 1976, p. 60; Hoyle, 1981, p. 77). In Williams' computer simulation described above, he decreased the summer sunshine substantially, corresponding to the astronomical theory of the ice age, but he still required a huge climate change to keep at least one inch of winter snow by the end of the summer just in northeast Canada. I lean toward the idea that scientists accept the astronomical theory more in desperation than for any other reason.

So, if the total duration of the ice age was about 700 years, it is very much less than the uniformitarian rate of 100,000 years for each ice age. The minimal erosion caused by ice sheets and the preservation of erosion surfaces in glaciated areas is further straightforward evidence for a rapid, post-Flood ice age (Lidmar-Bergström, Olsson, and Olvmo, 1997). Evidence indicates that only one ice age elapsed, not many.

This is only one example of the many processes thought to take too much time for the Scriptural time scale. These processes can happen rapidly using *different* assumptions, such as the Genesis Flood.

Figure 8.18 Three of a series of five dinosaur tracks (two badly eroded) forming a straight trackway on a bedding plane from northeast Wyoming.

HOW CAN DINOSAURS MAKE TRACKS AND LAY EGGS EARLY IN THE FLOOD?

There are a number of challenges for the time scale of the Flood based on sedimentary rocks and the fossils within those rocks. Some scientists will point out certain features of the rocks and fossils that they are convinced could not possibly have formed in a one-year Flood (Young, 1982; Wonderly, 1987; Strahler, 1987). Probably the most challenging example is the existence of dinosaur tracks and eggs found worldwide within what appears to be Flood rocks (Figures 8.16 and 8.17).

According to the Bible, all terrestrial, air-breathing animals died by Day 150 in the Flood (Genesis 7:19-24). Since tracks and eggs are activities of live dinosaurs, they must have been made early in the Flood, the Inundatory Stage of the Flood according to Walker's (1994) model. Critics charge that there is no way that dinosaurs could lay eggs and make tracks in the midst of a catastrophe as devastating as the early Flood, and there would not be enough time for that many tracks to be made. Since the tracks and eggs are found in sedimentary rocks, they would have to be assigned to the Flood. If they were made before the Flood, then they would have been destroyed by the Flood. Dinosaur tracks and eggs appear to be an insurmountable obstacle to the Flood model. So, some creationists have concluded the dinosaur tracks and eggs are a post-Flood feature (Garton, 1996; Garner, 1996).

Since I live in an area close to many dinosaur tracks and eggs, I decided to examine the area and literature on the subject (Oard, 1995, 1997c, 1998b). I discovered there are still many unknowns associated with the dinosaur data, and that there are a number of misconceptions in the literature concerning the eggs and tracks. John Horner had once thought that the worn teeth found in dinosaur babies were caused by the feeding of babies for a significant period of time by "good mothering lizards" (Horner, 1982). However, Horner and Phillip Curie later found embryos in dinosaur eggs that have worn teeth, leading to the conclusion that the dinosaur embryos ground their teeth *in the egg* (Horner and Currie, 1994). Horner and Dobb (1997, p. 153) point out:

> What about the worn teeth that we found first in the jaws of maiasaur nestlings, then in hypacrosaur embryos? It's now clear that dental wear is not evidence of parental care. My original interpretation turned out to be a misinterpretation...That leaves only one explanation: while inside their eggs, the embryos ground their diamond-shaped teeth together...

Worn teeth do not indicate a significant period of time as once thought.

The number of egg levels on Egg Mountain, Montana, is also questionable. It was stated by Horner (1982) that there were three egg levels, implying and a considerable amount of time between each horizon. These dinosaur-nesting horizons are likened to the nesting areas of certain birds. Although there are many whole eggs on these horizons and in the areas surrounding Egg Mountain, there is evidence that some eggs hatched. One piece of evidence is the dead juveniles around the nests. This would be another indication of too much elapsed time for the 150 days of the early Flood. However, the number of levels is in dispute (Oard, 1998b, pp. 73-76). Furthermore, Horner misidentified a rare embryo in one of the eggs on Egg Mountain, so the type of dinosaur that laid the eggs is *not* the same as the skeletons around the nest (Horner and Weishampel, 1996; Varricchio *et al.*, 1997). This leads to the conclusion that these eggs likely never hatched, and that the juvenile skeletons in the egg-laying area moved into the area and died where other dinosaurs had just laid

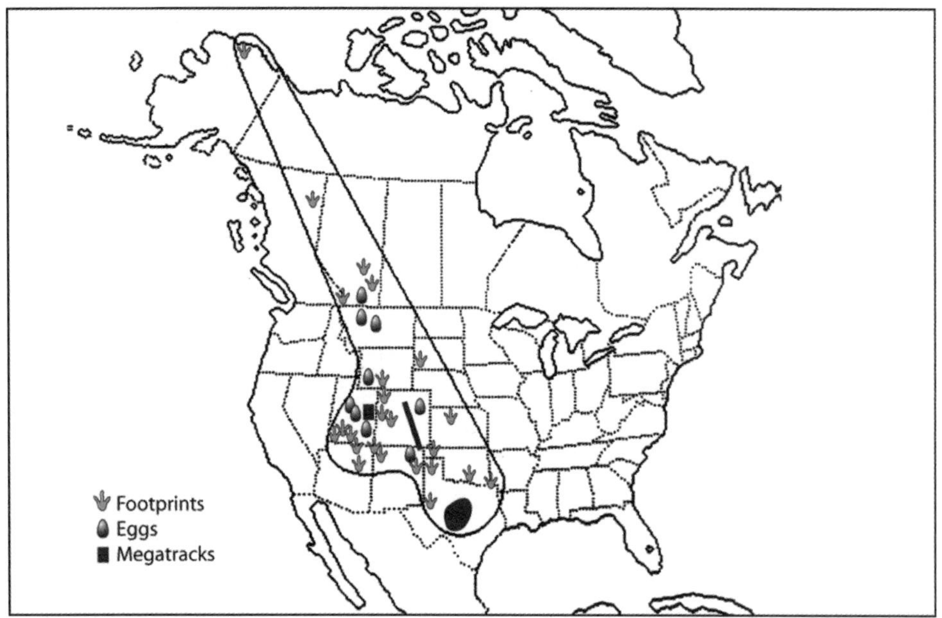

Figure 8.19 Area of exposed sediments or a series of shoals early in the Flood where dinosaurs could embark, make tracks, and lay eggs.

eggs. This need not take much time.

The above mistaken interpretations are listed in order to show the importance of using caution before jumping to unwarranted conclusions based on published data in the face of multiple unknowns.

When we examine tracks and eggs closer, we discover a number of puzzling features, if we assume they were made by normal dinosaur activity. The tracks are always found on horizontal or nearly horizontal bedding planes (Lockley and Hunt, 1995). Why wouldn't the tracks go from bed to bed? Were there not any hills to climb then? Practically all individual dinosaur trackways (multiple tracks made by one dinosaur) are straight (going in one direction) (Lockley, 1994). Figure 8.18 is a series of 5 dinosaur tracks, two badly eroded, that form a straight trackway on a bedding plane. Lockley and Hunt (1995, p. 165) write: "First, the sauropod was changing direction, turning to the right, a phenomenon *rarely* recorded in trackways" [emphasis mine]. Animals foraging for food, competing for mates, or other normal dinosaur activity commonly make curved, meandering, or sharply angled tracks. Straight tracks are usually made when an animal is fleeing something, such as a predator or hunter. During the catastrophic onslaught of the Flood, we would expect dinosaurs to flee the encroaching Floodwaters, making straight trackways (Oard, 2003)

It is unusual that there are very few tracks of babies and young juveniles (Lockley, 1991, pp. 31-32; Lockley and Hunt, 1995, pp. 121, 207). Modern track sites usually have a large percentage of baby and juvenile tracks. For instance, 50% of the elephant tracks in Amboseli National Park, Africa, were made by young elephants (Lockley, 1994, p. 359). There are very few tracks of dinosaurs that were most likely poor swimmers based on their morphology, such as stegosaurs, ankylosaurs, and ceratopsians (Lockley and Hunt, 1995, pp. 229, 231). None of these unusual characteristics of dinosaur tracks fit into the uniformitarian paradigm. The evidence better fits a time of worldwide stress for dinosaurs –a gigantic Flood.

Since the dinosaur tracks were made on hundreds to thousands of feet (500 meters or more) of Flood sediments, they indicate a period when sediments or shallow shoals were briefly exposed as the water was rising (Oard, 1995). Track layers on more than one bedding plane in an area can be explained as representative of brief exposures during a generally, continuous sedimentation event. This implies local sea level oscillations with sedimentation continuing during the next transgression, explaining how the tracks were covered rapidly enough for them to be preserved. Oscillations would be expected during a global Flood caused by tides, local and distant tectonic events, and the dynamics of Flood currents (Barnett and Baumgardner, 1994). Barnette and Baumgardner (1994), using computer modeling, found that huge currents could be generated on shallow continents, just by the spin of the earth or the Coriolis force. In the troughs of these currents, they found the sea level would fall and intersect the bottom. This pattern can remain stable for many days, leaving exposed land in the middle of a trough.

Figure 8.19 shows the suggested location for exposed Flood sediments or a series of shoals in the western United States early in the Flood. This would have been a depositional area very early in the Flood where thousands of feet of sediments were dumped. The deposition would have the effect of *shallowing* the area, making it more vulnerable to later exposure. A local fall in sea level due to any of the mechanisms already mentioned would have provided enough land for the dinosaurs to rapidly nest. They would have laid eggs and made tracks on the exposed land. Numerous bonebeds in this area have many unusual features –ones that are indicative of a watery catastrophe.

This exposed land before Day 150 in the Flood could also explain the nature of the dinosaur eggs and nests. It is natural to assume that many dinosaur females would be pregnant just before the Flood started. In a stressed condition they would lay their eggs and make nests on any temporary refuge they could find. A vivid illustration of this occurred in late February, 2000, in Mozambique (CNN news report, March 1, 2000). During a devastating flood, Cecilia Chirindza took refuge in a tree. As the flood surrounded her, she gave birth in the tree after three days of being stranded. Cecilia and the baby were rescued by a U.S. helicopter. When it was the baby's time, not even a flood could stop its birth.

The fact that very few nests are associated with the eggs indicates the eggs were laid in haste. They were often laid on simple bedding planes (Carpenter, Hirsch, and Horner, 1994). There are very few embryos associated with these eggs, also a possible indication of stressful conditions. Some of these eggs may have had time to hatch. The signs of hatching, however, such as eggs with broken tops, could have been caused by erosion when the next sedimentary layer was deposited, or compaction of the sediments, or the action of scavengers. There is evidence in egg laying areas and in bonebeds of scavengers, mainly in the form of carnivorous dinosaur teeth and teeth marks on bones.

In summary, there are a number of unusual features of dinosaur tracks and eggs that are more indicative of a stressful environment than a natural environment. It is reasonable to believe that newly deposited sediments could become exposed for a short time early in the Flood due to events occurring within the Flood. So, it is quite likely that all the dinosaur activity found in the sedimentary rocks did indeed happen during the first 150 days of the Genesis Flood.

Chapter 9

Missoula Flood Analogs For the Genesis Flood

Scientific literature does not give the impression that uniformitarian scientists have difficulty explaining the Earth's surface features. But if you dig a little deeper, you will find that not many features of the Earth's surface can be explained by uniformitarianism. For instance, mountain building has always been a mystery. Plate tectonics in association with subduction or convergence zones is the most popular explanation for mountains today. But, the theory has difficulty explaining mountains. Ollier and Pain (2000, p. 297) in a book on the origin of mountains summarize the difficulties: "A great many mountains, plateaus and other landscape features have no apparent relationship to plate tectonics situations." One persistent problem within the plate tectonics paradigm is finding room for magma "bubbling" up from a subduction zone and coagulating into a large mass of granite (Hutton *et al.*, 1990; Clemens and Mawer, 1992). For other difficulties of the plate tectonics paradigm see Reed (2000).

As developed in chapter 7, the origin of plains is also a paradox of geomorphology, especially the formation of planation surfaces. Other problems for uniformitarian theory are water and wind gaps. Presumably when the sediments in the valleys were thicker, the rivers could have easily flowed around these barriers, but instead "chose" to cut right through them. Continental shelves and submarine canyons are also difficult to explain. Many hypotheses have been devised to explain these landforms. However, all hypotheses have been shown to have considerable difficulty. It is amazing that after more than 100 years no single hypothesis for landform development is supported by a majority of geologists, as Crickmay (1974) noted in Chapter 7.

THE LIMITED USE OF ANALOGS

A key to unlocking the mysteries of geomorphology could lie in using analogs. An analog is loosely defined as a process or phenomenon that is similar to another process or phenomenon upon which a comparison may be made. A good, well-documented analog would be useful in providing insight into an unobserved process and aid in understanding the origin of the earth's surface features. It would also help in distinguishing between two or more competing hypotheses, such as answering the question of whether the surface landforms formed quickly and catastrophically or whether they took long ages. The Lake Missoula flood is one of the largest and most studied floods on the face of the Earth. It would be ideal for using as an analog for the Genesis Flood. There is a scale problem, however–the Genesis Flood was very much larger than the Lake Missoula flood. As a result, other unique phenomena, which no analog can provide, are expected.

The danger of taking analogs as exact representations is demonstrated by those who attempted to use modern ice-burst dams to constrain the flow from glacial Lake Missoula

Figure 9.1 Cross-bedded sandstones from Checkerboard Mesa, Zion National Park, Utah.

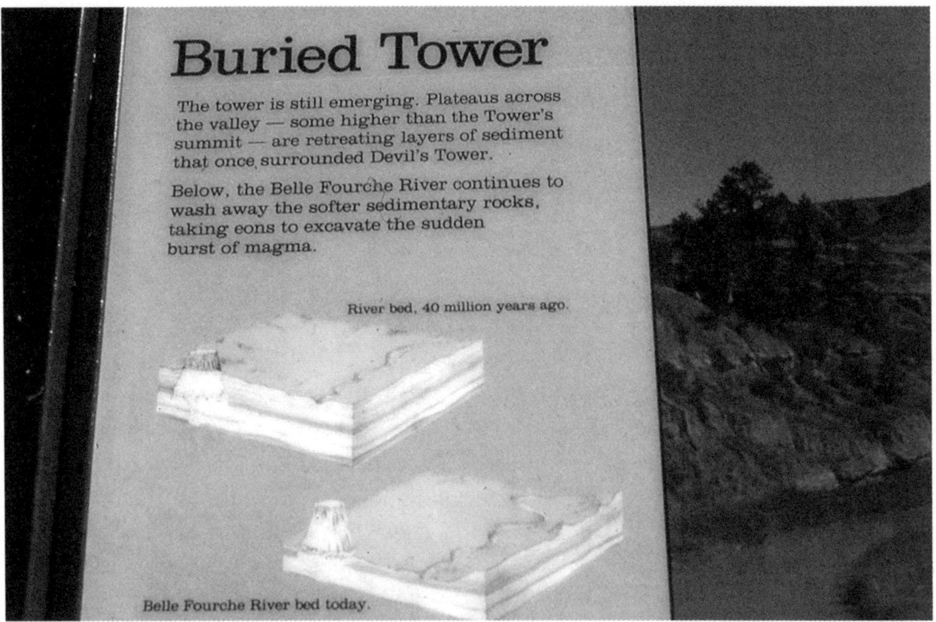

Figure 9.2 Road sign at Devils Tower explaining that erosion exposed the top of the tower about 40 million years ago in uniformitarian time. Erosion since then has leveled the surrounding plains to the present without hardly touching the tower, although it stands tall and lean. This is unreasonable considering modern, observed erosion rates. It seems that only rapid sheet erosion that did not completely scour the hardest remnants is required to explain Devils Tower.

(see Chapter 3). This is a typical example of the scale problem. By using analogs of modern ice dam breaks that were tiny in comparison to the Lake Missoula flood, scientists made significant mistakes in estimating the volume of water released at the dam site. In one case they estimated 10% the volume of the actual flood. So also the same "scale problem" exists when comparing the Lake Missoula flood to the much larger Genesis Flood. Regardless, we can gain some insight into the origin of some rock formations or landforms by comparison with analogs.

INUNDATORY STAGE ANALOGS

Our main focus will be on analogs for the Recessional Stage of the Flood, but using the Lake Missoula flood we can also find a crude analog for the Inundatory Stage of the Genesis Flood. I will be applying Walker's Biblical geological model. During the Lake Missoula flood, 50 mi^3 (200 km^3) of silt and basalt was eroded from eastern Washington in a few days. The Genesis Flood had a volume of water possibly a million times the volume of the Lake Missoula flood. This would lead us to expect a gigantic amount of erosion to occur within a short amount of time from the power of flowing water, the tectonic disruption of the land, volcanic input, debris from meteorite impacts, and other processes acting during the Flood.

Sediments from the Genesis Flood would have been deposited mostly in areas where the currents slowed. The Walla Walla slackwater rhythmites, as exemplified in Burlingame Canyon with its deposition of 39 graded beds that were laid down in about one week, is analogous to the sedimentary rocks from the Genesis Flood that cover 75% of the Earth surface (Figure 9.1). Great gravel bars, such as the Ephrata Fan and the Portland Delta, were deposited within just a few days. A violent Flood of one year could easily deposit thousands of feet (many hundreds of meters) of sedimentary layers in "slackwater" areas within a very short time.

The Lake Missoula flood is an imperfect example for the Inundatory Stage of the Flood, but it teaches us the erosive power of moving water. It also demonstrates that all of the sedimentary rocks we see could have been deposited rapidly as layers, contrary to the idea of one gigantic chaos of undifferentiated mud as some have wrongly envisioned. Lastly, the Lake Missoula flood communicates to us that the Genesis Flood surely could not have been a "tranquil flood."

Figure 9.3 Close up of Devils Tower showing the strongly jointed nature of the rock.

TALL, LEAN EROSIONAL REMNANTS

We can gain insight into the Recessional Stage of the Flood by considering the erosional remnants of the Lake Missoula Flood, especially the high towers of rock it left behind or failed to erode.

In Chapter 7, I discussed several examples of tall, thin erosional remnants of igneous rocks left from the Genesis Flood. Devils Tower was an example (see Figure 7.6). Since Devils Tower is the throat of a volcano, the Great Plains sedimentary rocks at one time had to have been as high as or higher than the top of the tower. The cone of the volcano would have been higher than the tower and covered the sedimentary rocks surrounding the volcanic neck. There is a road sign at Devils Tower that illustrates how it allegedly remained standing even as erosion carried away the sedimentary rocks around it (Figure 9.2). The sign asks us to believe that the tower itself stood for 40 million or more years while this slow process of erosion did its work. This is not reasonable. Why should Devils Tower be left hanging for 40 million years or more and yet show comparatively little erosion? It is especially difficult to justify all this time when we consider North America would be eroded to sea level in less than 50 million years from present erosional processes. Granted, Devils Tower is more resistant to erosion, being an igneous rock, but it would still crumble into a pile of rubble within a relatively short time, because it is similar to basalt and is well jointed (Figure 9.3). Water easily penetrates joints allowing frost action to break it to pieces. Freeze-thaw cycles are quite common in this part of the United States. It seems that the only logical mechanism that would leave such an erosional remnant is a sheet of water rapidly washing away the plains sedimentary rocks, leaving behind the harder erosional remnants. This implies rapid sheet erosion. It is likely that if sheet erosion continued any longer, it would also have completely destroyed Devils Tower. Are there any analogs for such erosional remnants from the Lake Missoula flood? The best example is Steamboat Rock in the Upper Grand Coulee.

Figure 9.4 Beacon Rock in the Columbia Gorge viewed eastward from Cape Horn Viewpoint, Washington.

Figure 9.5 Small erosional remnant next to basalt cliffs just downstream from Wallula Gap, Washington.

This one mi^2 (2.6 km^2) rock lies in the middle of the upper end of the coulee and is about 900 feet (275 meters) tall, the same height as the coulee walls. This rock was part of the continuous basalt cover that once filled the area now occupied by the Upper Grand Coulee. The Lake Missoula flood eroded all around Steamboat Rock, leaving it as an isolated remnant. As with Devils Tower, if the Lake Missoula flood continued or there was more than one gigantic flood, Steamboat Rock likely would have disappeared by erosion.

Steamboat Rock works better as an analog for the formation of the monuments in Monument Valley (see Figure 7.8) during the more channelized flow of the Recessional

Stage. Monument Valley is a north-south broad valley about 30 miles (50 kilometers) wide with a flat floor carved on a broad anticline. The carving of the valley left monuments as high as 1000 feet (305 meters). The monuments are capped by sandstone that is soft enough to be easily rubbed off using your fingers.

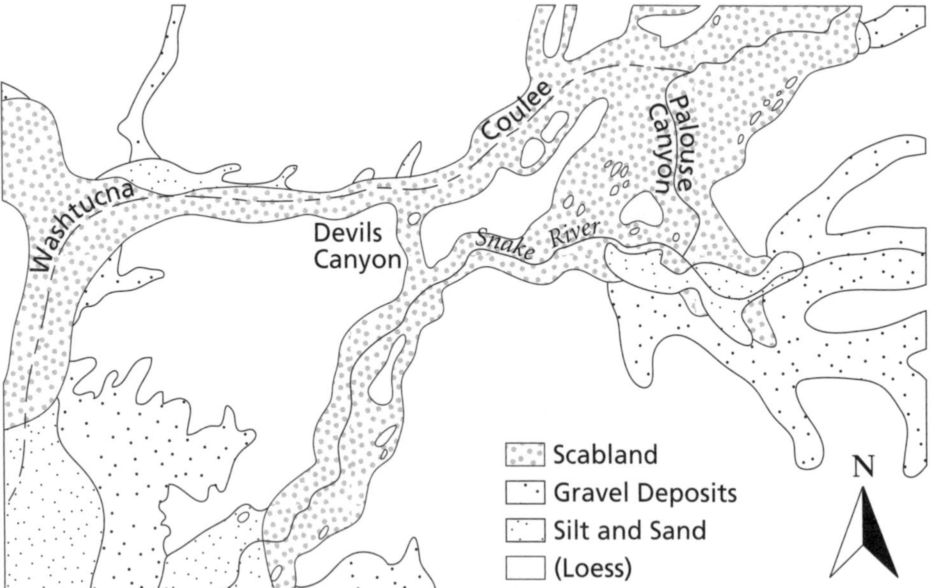

Figure 9.6 Map of ridge between Washtucna Coulee and Snake River (redrawn from Bretz, 1928a, p. 205 by Mark Wolfe).

Beacon Rock, like Devils Tower, is also the throat of a volcano that is 848 feet (259 meters) high. It lies in the middle of the Columbia Gorge (Figure 9.4). Although the Lake Missoula flood raced down the Columbia Gorge at up to 80 mph (35 m/sec) with a large number of boulders bouncing along the bottom, this remnant stood. The boulders would have impacted the base of Beacon Rock and, if the flood lasted longer or there were dozens of floods, the moving boulders likely would have destroyed Beacon Rock.

There are many smaller erosional remnants of basalt (Figure 9.5) and streamlined silt hills in the flood path. These analogs from the Lake Missoula flood make it more reasonable to believe that the Genesis Flood could leave the tall, lean erosional remnants after rapid sheet erosion.

Figure 9.7 Palouse Falls within Palouse Canyon, southeast Washington.

CONTINENTAL SHELVES AND SUBMARINE CANYONS

Continental shelves and submarine canyons are further paradoxes of geomorphology that also find a crude analog in the Lake Missoula flood.

Continental shelves have the appearance of having been first deposited rapidly by sheet deposition and secondly cut by channelized erosion to form submarine canyons during the Recessional Stage of the Genesis Flood. Expansion bars that developed during the Lake Missoula flood provide a rough analog for the rapid formation of continental shelves with their submarine canyons. The most helpful analog is the Portland Delta (see Figure 2.21). Another is the Ephrata Fan. Both were deposited as water exited narrow constrictions.

After most or all of the sediment was deposited, channels were cut on top of both the Portland Delta and the Ephrata Fan. The more channelized water that exited the Columbia Gorge and the Willamette Valley eroded deep channels through the Portland Delta. Bretz (1928d, p. 697) concluded that the current Columbia River did *not* cut its channel through the Portland Delta, but the waning phase of the Lake Missoula flood did the excavating: "The broad channel of the river between the two parts of the delta is interpreted as essentially the Spokane Flood channel not as a product of later dissection [by the Columbia River]." The Columbia River, of course, would naturally flow through this channel after the flood. Likewise, the water exiting the Willamette Valley cut a channel in the debris along the east edge of the Portland Hills, an anticline in the Columbia River Basalt. Since the channel is the lowest point, the Willamette River eventually flowed through it. Both of these channels are crude analogs for the formation of submarine canyons. Shallow channels were also cut on top of the Ephrata Fan during the waning, or channelized flow of the Lake Missoula flood (Bretz, 1959, pp. 32-33). Long, narrow Moses Lake occupies three channels.

Figure 9.8 The blocked entrance to Devils Coulee (view southwest from Washtucna Coulee).

The Missoula Flood is not an exact analog for the formation of continental shelves and submarine canyons because the debris in the Portland Delta and Ephrata Fan is much coarser with downstream dipping foreset beds, while the continental shelves are composed of finer-grained debris, such as sand, silt, and clay, with generally planar sediments. The difference has several possible causes. It could be related to the distance of transport. For instance, the cobble and boulder debris deposited on the Portland Delta and the Ephrata Fan were transported only a short distance from their upstream, constricted gaps. Debris in the Genesis Flood flowed a long distance from off the continent to the continental margin. Such a long transport distance would have decreased the size of the particles in the sediment. In rivers, it is well known that fine sediment travels downstream farther than coarse sediment. Another factor could be the initial size of the debris.

WATER AND WIND GAPS

Another mysterious feature, discussed in Chapter 7, and commonly found worldwide is water and wind gaps. It is interesting that the Lake Missoula flood provides almost an exact analog for the rapid formation of water and wind gaps by a gigantic flood.

Washtucna Coulee is an east-west coulee that lies generally parallel to and about 10 miles (16 kilometers) north of the Snake River (Figure 9.6). The basalt ridge between the valleys was breached in two places by the Lake Missoula flood. One breach became a water gap and the other a wind gap, called Palouse Canyon and Devils Coulee, respectively.

Each gap was cut 500 feet (150 meters) deep through the ridge within a matter of days. After the flood the Palouse River that formerly flowed westward down Washtucna Coulee and eventually into the Columbia River was diverted 90^0 to the south to flow across the transverse barrier through Palouse Canyon into the Snake River after the flood ceased (Bretz, 1928a). The beautiful Palouse Falls can be seen in this section (Figure 9.7).

Devils Coulee developed into a wind gap. It was cut through the ridge 15 miles west of Palouse Canyon. The Lake Missoula flood did not cut the entrance from Washtucna Coulee deep enough. A rock sill 100 feet above Washtucna Coulee blocks the entrance to Devils Coulee (Figure 9.8). This sill keeps any water flowing in Washtucna coulee from taking a 90^0 turn into Devils Coulee to the Snake River.

If a geologist did not know about the Lake Missoula flood, he would postulate one of the three main uniformitarian hypotheses for the formation of these water and wind gaps: antecedence, superposition, or stream piracy. But the water and wind gaps formed catastrophically when water from the Lake Missoula flood overtopped the ridge. In a similar way, water and wind gaps from around the world can easily be explained by the Genesis Flood rapidly carving out these gaps, starting with water overtopping the transverse ridges.

Chapter 10

The Meaning of the Controversy for Today

Although the number of Lake Missoula floods is the main debate today, geologists have settled long ago on the fact of at least one colossal Lake Missoula flood. The controversy and acceptance of the flood is touted as one more example of how science eventually finds the truth. However, there is another story to tell. The controversy tells of scientific prejudices in the face of obvious evidence for a flood. The Lake Missoula flood may have been accepted 40 years ago, but the attitude behind its delayed acceptance lives on. Yes, the same old prejudices that blinded the geologists are still with us in force today and affect many aspects of our lives. I will recall events of the Lake Missoula flood controversy in developing this theme.

SUMMERY OF THE LAKE MISSOULA FLOOD CONTROVERSY

J Harland Bretz spent years studying the geology of eastern Washington. In his explorations he noticed dry waterfalls, exotic boulders, a plexus of youthful canyons, water-shaped silt hills, and giant gravel bars. He came to the inescapable and vastly unpopular conclusion that a gigantic flood several hundred feet deep had washed across the entire area in fairly recent geological time. At first he did not understand the source of all the water, but later he connected the huge ice-dammed lake in western Montana with the flood. The rejection of Bretz's flood spanned 40 years of contention and controversy among the geologists of the time. They ignored or minimized his observations, often refusing to *examine evidence in the field*, and denied the reality of such a flood. They even invented other "outrageous hypotheses" to counter Bretz's claim because Bretz's conclusions smacked too much of the biblical Flood.

The weight of the observations eventually brought acceptance of Bretz's theory. Then geologists started thinking they found pockets of evidence for, at first, several floods, then 40 floods, and finally about 100 floods during the last ice age. Lately, some investigators reexamined the evidence and came to the conclusion that there was only one colossal Lake Missoula flood with perhaps several minor floods. This likely is the correct conclusion, although most scientists lean more towards multiple floods. Once the Lake Missoula flood was established, scientists recognized evidence for similar floods in other parts of the Northern Hemisphere, the most interesting are the immense subglacial floods suggested by John Shaw and colleagues. Bretz was able to open their eyes to a whole new world of catastrophic floods. Surely, scientists have learned their lesson to not cling too tightly to uniformitarianism. Or have they?

THE UNSHAKEABLE FAITH AGAINST THE MISSOULA FLOOD

Bretz encountered intense opposition almost immediately, especially during the 1927 showdown at the Geological Society in Washington, D.C. (Baker, 1978a, pp. 7-9). At the meeting, they strongly challenged him and proposed an array of bewildering hypotheses to explain the Channeled Scabland. These other "outrageous hypotheses" became more sophisticated in the 1930s when Allison and Flint proposed a giant blockage of the Columbia River in the Columbia Gorge that backed up into eastern Washington. Although both men had examined the field evidence in depth, in the end they saw *only* slow gradual processes over millions of years. What examples of the power of preconceived opinion! With such a bias even fieldwork is no real help in discovering the truth. Baker (1978a, pp. 11-12) attests to how such preconceived convictions can blind a geologist to the obvious:

> Bretz's flood theory was so despicable that even circular reasoning could be employed to erect an alternative hypothesis.
>
> A careful examination of Flint's...paper reveals that he observed and described the morphological feature which, more than any other, was absolutely incompatible with his elegant theory.

Remember that Flint was considered the patriarch of glacial geology of his time.

The strong bias against catastrophism shows up again when Joseph Pardee discovered the "inland sea" or Great Lake of glacial Lake Missoula (Alt, 2001, pp. 7-10) but failed to take his research to its logical conclusion as the cause of the Channeled Scabland. When Pardee was present at the 1927 Washington, D.C., debate with Bretz, it is reported that he turned to Kirk Bryan and said, "I know where Bretz's flood came from" (Baker, 1978a, p. 9). Why did he not speak up in defense of Bretz? There is proof that even at that time,

Pardee was considering the idea that glacial Lake Missoula carved the scablands, but his supervisor, W.C. Alden, dissuaded him (Baker, 1978a, p. 8). Sometimes truly great ideas are suppressed from above, and those who develop them are too afraid to speak up for what they believe. Pardee retired from the U.S. Geological Survey in 1940. In that year he presented a paper about "unusual currents" in glacial Lake Missoula (Pardee, 1940). The paper was published in 1942. Pardee (1942) described the Camas Prairie ripple marks and other features demonstrating catastrophic currents. His title is quite innocuous and does not give the reader much insight into the content of the paper. The paper is filled with weasel words, such as "The ice dam is thought to have failed." Pardee obviously did not desire to rock the geological boat anymore than necessary, even after he retired. His paper was an understatement throughout, and he never did state the obvious connection between the flood and the Channeled Scabland. Maybe he hoped Bretz would make the connection and so avoid aligning himself with an unpopular hypothesis (Allen, Burns, and Sargent, 1986, p. 66). Regardless, few grasped the significance of Pardee's innocuous paper at the time.

Now that the controversy has died down and Bretz is a hero, Baker can announce to the world that the Lake Missoula flood controversy is yet another "...marvelous exposition of the scientific method" (Baker, 1978a, p. 3). Error was finally discovered and repaired just like with the faked Piltdown Man and the grossly misdiagnosed Neandertal Man (Lubenow, 1992). It is interesting how scientists can use such revealing examples of blindness as lessons in the scientific method. I see a deeper meaning in the flap over the Lake Missoula flood.

FAITH THAT THERE WAS NO GLOBAL FLOOD

In a like manner, there is strong evidence for a worldwide Flood in the form of vast sedimentary blankets, billions of fossils, tall erosional remnants, continental shelves, submarine canyons, planation surfaces, water gaps, wind gaps, etc. This evidence continues to be ignored, suppressed, or explained away by unreasonable hypotheses. Furthermore, the mechanism of the Flood at this point is unknown, although there are several proposals. Just as Bretz did not know the cause of the Lake Missoula flood for 10 years, the lack of a Flood mechanism does not mean the Flood never occurred.

Uniformitarianism continues to lock in concrete the thinking of scientists today, and yet it is a mere assumption. Geologists are rarely able to interpret the rocks in a way that varies from their uniformitarian framework. David Alt (2001, p. 181) comments on the blindness and stubbornness of mainstream geologists towards the Lake Missoula flood–an attitude that continues to the present day:

> It would be pleasant to write that Bretz's detractors saw the light of evidence and reason, and changed their minds. Very few did...Most of Bretz's critics went to their graves still stoutly maintaining that perfectly ordinary processes of erosion had somehow produced the extraordinary landscapes of the scablands...It seems customary for those who tell it to finish with a rousing homily about how this episode opened the eyes of geologists to the *great world of catastrophic events*, inspired them to explore *new* avenues in their thinking. I wish that were so, but know that it is not [italics mine].

Scientists in general and geologists in particular appear *not* to have learned anything from the Missoula flood controversy. They continue to censor ideas that the geological establishment finds disagreeable. Any hint of catastrophism is still suppressed by most scientists. For instance, John Shaw's subglacial flood hypothesis is considered an outrageous hypothesis today. Listen to some of his complaints in attempting to publish his ideas:

> Over the period of the recent summer Olympics, I broke a personal record, zero for four on submitted manuscripts...Each paper challenges conventional wisdom on drumlin formation and was rejected by drumlin "experts" who have contributed to that wisdom...There is a potential for conflict of interest here, and I am convinced that it exists when the negative reviews, which call for outright rejection, are lacking in substance, insulting, and commonly sarcastic (Shaw, 1988b, p. 291).

This treatment is in spite of the fact that the origin of drumlins is yet to be determined. Shaw (2002, pp. 19-20) states in a later article how some scientists condemned the subglacial flood hypothesis outright, that there has been much politicking behind the scenes, and that graduate students who chose to work on outburst flood landforms have been through difficult times.

The controversy surrounding the Lake Missoula flood demonstrates that even if evidence for a worldwide Flood were obvious, as it is in many places, the scientific establishment would refuse to see it. And even for the few who can see such evidence, their conclusions will often be suppressed by the intellectual establishment, as happened to Joseph Pardee. What Alt (2001, p. 126) wrote in regard to the Lake Missoula flood has application here: "Of course, it is easily possible to be so wrong if you make up your mind without having considered or even seen the evidence." Most scientists are *not* open minded or objective at all if the data go against accepted assumptions, especially if someone provides evidence for the possibility of a global Flood. Even Victor Baker (1978d, p. 1255), one of the early defenders of one Lake Missoula flood, demonstrates his unwillingness to entertain any evidence for a global catastrophe:

> The catastrophist idea of the Noachian debacle was finally laid to rest when Louis Agassiz showed that

his glacial theory could explain erratics, striations, till, fluvioglacial features, and so on. Old ideas die hard, however, and catastrophist absurdities still appeared in the literature of the early 1900's (as they do even today).

The surficial "diluvium" was the last bastion for the Genesis Flood that some geologists defended in the early 1800's. Unfortunately, the scientists of that time were operating under wrong preconceived ideas. The surficial diluvium is from the post-Flood ice age (Oard, 1990), while the hard sedimentary rocks underneath with billions of fossils are the remains of the real Genesis Flood. When Agassiz's glacial theory showed that the diluvium was glacial debris in the mid 1800's, geologists had no Genesis Flood left. Geologists today believe the question of the Flood was settled back then, but that was at a time when little geology was known.

The Genesis Flood was never disproved, only assumed out of existence in the minds of geologists since the early 1800s (Whitcomb and Morris, 1961). Based on the biblical geological model of Tas Walker (1994), rock features and fossils line up much better with the Genesis Flood and do not fit with uniformitarianism, the "slow processes over millions of years" assumption. The short time scale of Scripture at first appears to threaten all gains made in understanding the biblical Flood. But, as we consider when and why scientists introduced the concept of an old earth, we find this concept began as an *assumption* in the mid 1700s. The old earth concept actually predates uniformitarianism and evolution and was offered because some scholars of the day rejected the global Flood. Radiometric dating, discovered in the 20th century, produced very old dates. Because of the old dates, radiometric dating was readily accepted, in spite of the fact that it is based on many poor assumptions and is therefore unreliable. Most dating methods give younger ages than radiometric dating. Planation surfaces are often dated by radiometric techniques and fossils as being tens of millions to over 100 million years old, but if that is true, they should have eroded away long ago. There are other geological processes that are thought to have taken too much time for the short chronology of Scripture. The Pleistocene ice age is a good example. With different initial conditions, this ice age would have developed after the Flood. An ice age would have taken approximately 700 years to develop and melt, much less time than an ice age based on uniformitarian assumptions. Dinosaur tracks and eggs at first appear to require too much time than the year of the Flood, but they can be explained as having been made early in the Flood. The Lake Missoula flood debacle is very useful to creationists not only because it gives us insight into how uniformitarian blinders affect the interpretation of data, but also gives us excellent clues as to how water and wind gaps, and other difficult-to-explain geologic features, could have formed rapidly during a global Flood.

THE MISSOULA FLOOD–ONLY THE TIP OF THE ICEBERG

Does such intense opposition to the global Flood, exemplified in the Missoula flood controversy, spill over into other areas of science, as well as to life? Why should it matter to the Church or to us individually?

We need to keep in mind that the Lake Missoula flood controversy and the opposition to a global Flood display an *attitude* or *worldview* that originates from an evolutionary-uniformitarian basis. This is just not an isolated incident without meaning for today. This worldview was mainly developed for those that wanted an alternative to biblical Christianity.

Most people are unaware of the history of this worldview or its purpose, yet today it greatly impacts our society. The paradigm has all the earmarks of a religious belief system because it is not based on provable evidence or scientific fact but on faith. It answers the questions of who I am, what is my purpose, what do I believe about the universe, and where am I going after I die. As a result it affects society's values and our individual choices, since the worldview dominates our intellectual institutions and therefore, strongly influence what people today think and believe. The evolutionary-uniformitarian paradigm is a system that automatically excludes God as a reasonable hypothesis or expectation for the prehistoric, unobservable past. It defines man as an evolved animal. It assumes *a priori* that matter is all there is (naturalism or materialism), and that all factual or reasonable (scientific) explanations must be interpreted within natural processes only. It impacts our attitude towards ourselves and others. Finally, it answers the question of who is in charge: God or man?

The evolutionary worldview with its underlying philosophy of naturalism or materialism (Johnson, 1991) dominates the school system and is reinforced in museums, national parks, nature shows, and other television programs. It is taught, sometimes subtly, to children beginning in kindergarten. To oppose this worldview brings fierce opposition and threats of lawsuits. The fruit of this worldview, of which the Lake Missoula flood is only the tip of the iceberg, is that few school age children are being taught in school that there are absolutes, except that evolution is a "fact." It follows that morality is relative, if we are only evolved animals. The Ten Commandments are "out," leaving institutions and values of society that depend upon opinion polls. Marriage and the family are no longer upheld as sacred or necessarily desirable. Today, we see western society moving away from the Judeo-Christian worldview. Nearly every social indicator has shown greater problems since about the 1960s.

Few remember how the evolutionary-uniformitarian paradigm affected sociology and the philosophy of government. Social Darwinism provided the world a "scientific" basis for racism (Haller, 1971; Ham, Wieland, and Batten, 1999; Bergman, 2001, 2002). Survival of the fittest as an alternative to Christianity allowed the state to replace God in communism. This was also the scientific underpinning of Hitler eliminating the "weak and unfit" (Barzun, 1958; Himmelfarb, 1962, pp. 412-431; Morris, 1989; Perloff, 1999, pp. 219-235). Historian Jacques Barzun (1958, pp. 5-6) stated: "The results of making matter the only reality was plain: a premium was put on fact, brute force, valueless existence, and bare survival." As the movers and shakers of society continue to apply the evolutionary-uniformitarian paradigm, it will be a slow march towards Armageddon.

Unfortunately, acceptance of evolution has crept into the Church. At first in the mid and late1800s it was barely noticeable, but it has led to much compromise. To allow for the acceptance of evolution, Genesis had to be relegated to myth or allegory, or "reinterpreted." Questioning scripture birthed the school of higher criticism. Their ideas are solidly entrenched in many mainline churches. The Jesus Seminar is just one fruit. This "seminar" is made up of a large group of "theologians" that gathered together to "vote" on what really were the sayings of Jesus in the New Testament. Amazingly, the seminar participants voted that most statements made by Jesus were not spoken by Him but added later by his disciples. Little of the book of John remained as authentically from Jesus. Even the science and theological departments in most of the evangelical, fundamentalist, and charismatic colleges have gradually fallen to accepting the evolutionary-uniformitarian paradigm, mainly because it appears intellectual. This has led to special "interpretations" of the Bible that sap the vitality of these institutions. Compromising Scripture has led to liberalism, the rejection of the biblical faith by millions of people, and to the irrelevance of the Church in today's society.

The Lake Missoula flood controversy highlights the worldview problem. In this study we see how deeply flawed the research and conclusions can be when interpreting the past. This prejudice should alert us to other wrong interpretations and cause us to examine carefully what they purport as truth. Naturalism's foundation is built on the unobservable and unprovable past. It is important to remember that we should never compromise the Word of God for the sake of man's speculations.

There is a whole new world of discovery waiting for those who throw off the evolutionary-uniformitarian worldview and accept God's word as the final arbitrator. The future is very bright and exciting for a discerning Christian and especially for a creation scientist who is willing to take the time to carefully reexamine the raw data. Today there are mountains of data to comb through and reinterpret. As Kepler once said, "Science is discovering God's laws after him." A careful study of Scripture and science will reveal their agreement because they both have the same Author.

Glossary

Analog–a process or phenomenon that is similar to another process or phenomenon upon which a comparison may be made.

Antecedent river–a river that continued to flow and erode over the same location while a mountain barrier slowly lifted upward across its path.

Anticline–a fold in the rock layers, generally convex upward, whose core contains stratigraphically older rocks.

Argillite–a compact rock derived from mudstone or shale that has been weakly metamorphosed and further hardened.

Astronomical Unit–the distance between the sun and earth.

Basalt–a dark-colored volcanic rock with a high proportion of magnesium and iron

Base level–the theoretical limit or lowest level toward which erosion progresses but seldom reaches. Sea level is considered the ultimate base level.

Boulder–a detached rock mass larger than a cobble having a diameter greater than 10 inches (25.6 centimeters).

Butte and basin topography–landscape characterized by small mesas and basins with vertical cliffs.

Catastrophism–the doctrine that sudden violent, short-lived, more or less worldwide events outside our experience have greatly modified the earth's crust.

Cavitation–the explosion of bubbles against rock caused by high speed flow in shallow water.

Cobble–a rock fragment having a diameter between 2.5 and 10 inches (6.4 and 2.56 centimeters).

Cordilleran Ice Sheet–the ice sheet that covered British Columbia, the adjacent Alberta Rocky Mountains, northern Washington, northern Idaho, and northwest Montana during the ice age.

Coulee–in the northwest U.S. a generally dry trench-like valley.

Daughter element–the final stable element or isotope of an element of radioactive decay.

Debris flow–a moving mass of rock fragments of all sizes within a finer-grained matrix.

Debrites–the deposited product of a debris flow.

Driftless area–area within the boundary of an ice sheet that was not glaciated.

Drumlin–rounded or elongated hill composed of till or stratified glaciofluvial sediments associated with the ice age.

Duricrust–a general term for a hard crust on the surface, or a layer in the upper horizons of a soil in a semiarid climate.

Eddy bar–a bar formed at the mouth of a tributary valley or in a small valley high up on the side of a channel due to two converging flows or a swirling eddy.

Erratic or exotic boulder–a boulder that lies on terrain with a lithology different from the boulder.

Esker–a linear, winding ridge of coarse stratified gravel and sand left behind by a stream flowing within or beneath stagnant ice.

Expansion bar–a bar developed as the flow of water spreads out and slows after passing through a narrow constriction.

Flute–a scoop-shaped depression or groove.

Fluvial–of or pertaining to a stream or river.

Foreset bedding–beds of gravel that dip downflow, like a gravel delta,

Geomorphology–the science that treats the general configuration of the earth's surface.

Hogback–narrow ridges of steeply dipping rock layers.

Interlobate stratified moraine–a moraine between glacial lobes that contain stratified glacial debris.

Jökulhlaup–the flood caused by the bursting of an ice-dammed lake.

Kolk–a very strong vertical vortex that develops within deep, super-fast flowing water. It is similar to a tornado but is underwater

Lahar–a volcanic debris flow.

Laminations–a random alternation of two or more thin sedimentary layers.

Landform–one of the many features taken together make up the surface of the earth.

Lateral moraine–a moraine, which is a mound or ridge of glacial debris, that forms along the edge of a glacier.

Laurentide Ice Sheet–the ice sheet that covered practically all of Canada east of the Rocky Mountains and the northern United States during the ice age.

Lava tube–a hallow space beneath the surface of a solidified lava flow, formed by the withdrawal of molten lava after the surface solidified.

Loess–a blanket deposit of mostly silt that is deposited by the wind.

Materialism–the doctrine or philosophy that matter is the only reality, and that everything in the universe, including thoughts, the will, and feelings, can be explained only in terms of matter without recourse to the supernatural. A synonym for naturalism.

Metamorphic rock–any rock derived from pre-existing rocks by mineralogical, chemical, and/or structural changes.

Monocline–a local steepening of otherwise horizontal or uniformly dipping rock layers.

Naturalism–the doctrine or philosophy that matter is the only reality, and that everything in the universe, including thoughts, the will, and feelings, can be explained only in terms of matter without recourse to the supernatural. A synonym for materialism.

Orogeny–the process of formation of mountains.

Paradigm–a supermodel, worldview, or way of looking at a set of data.

Parent radioactive element–the beginning element or isotope of an element that decays radioactively into another element or isotope.

Pendant or longitudinal bar–a bar that forms downstream from an obstacle.

Peneplain–a low, nearly featureless, gently rolling land surface of considerable area that presumably has been produced by the processes of Davis's "cycle of erosion."

Percussion mark–a semicircular crack on a rock.

Planation or erosion surface–a land surface shaped and subdued by the action of erosion, especially by running water. The term is generally applied to a level or nearly level surface. An erosion surface can be more rolling while a planation surface is nearly level.

Pleistocene–an epoch of geological time from about 2 million years ago to 10,000 years ago in the uniformitarian time scale. It is generally associated with the ice age.

Pliocene–an epoch of geological time just before the Pleistocene.

Pluvial lake–a lake formed during a period of high precipitation.

Pothole–a smooth deep bowl-shaped or cylindrical hollow formed in a stream by the grinding action of stones or coarse sediment being whirled around by an eddy. Potholes can be formed by meltwater in glaciated areas.

Rogen moraine–small ridges and troughs of debris transverse to glacial or glacial floodwater motion.

Rhythmite–a repeating vertical sequence of two or more sediment types in a particular order.

Scabland–an area underlain by basalt with a thin soil cover and sparse vegetation and usually with deep, dry channels scoured into the surface.

Sichelwannen–a small sickle-shaped erosional depression with a sharp upstream rim.

Strandline–the ephemeral line or level at which a body of standing water meets the land. It can refer to a shoreline or a former shoreline now elevated above the present water level.

Stream piracy–the idea that one stream erodes faster than another and eventually captures the headwaters of the second stream.

Striated pavement–scratches or grooves on a bedrock surface.

Striated rock–scratches or grooves on a rock, usually a cobble or boulder.

Superposition–the hypothesis that a drainage pattern was developed while the whole land was nearly flat, so that the drainage was superimposed downward onto the current topography during erosion.

Tectonics–a branch of geology dealing with the broad architecture of the outer part of the earth, mainly the major structural or deformational features and their origin.

Tephra–material ejected from a volcano through the air.

Tephrochronology–the dating method that uses various chemical and other signatures to arrive at a date for volcanic ash.

Terminal or end moraine–a moraine, which is a mound or ridge of glacial debris, that forms at the terminal position of a glacier.

Tertiary–the first period of the Cenozoic era from 65 million to 2 million years in the uniformitarian time scale.

Till–debris consisting of stones inside a fine-grained matrix that was deposited by a glacier or ice sheet.

Tillite–a consolidated till.

Tunnel valley–a shallow trench cut in glacial debris or bedrock by a subglacial stream.

Turbidite–the deposit of a turbidity current.

Turbidity current–a density current in water or air, mainly a bottom-flowing current laden with suspended sediment that moves swiftly down a subaqueous slope and spreads horizontally on the floor of the water body.

Uniformitarianism–the fundamental principle that geological processes and natural laws now operating to modify the earth's crust have acted in much the same manner and with essentially the same intensity throughout geological time. It is easily remembered by the classical concept that "the present is the key to the past."

Varve–a rhythmite, generally composed of a silt/clay couplet, deposited in a lake during a one-year period.

Water gap–a deep pass in a mountain ridge, through which a stream flows; esp. a narrow gorge or ravine cut through resistant rocks.

Wind gap–a shallow notch in the crest or upper part of a mountain ridge, usually at a higher level than a water gap

Bibliography

Ager, D.V., 1973. *The Nature of the Stratigraphic Record*, Macmillan, London.

Alden, W.C., 1932. *Physiography and Glacial Geology of Eastern Montana and Adjacent Areas.* U.S. Geological Survey Professional Paper 174, U.S. Government Printing Office, Washington, D.C.

Allen, J.E. 1991. The case of the inverted auriferous paleo-torrent– exotic quartzite gravels on the Wallowa Mountain peaks. *Oregon Geology* 43:104-107.

Allen, J.E., Burns, M., and Sargent, S.C., 1986. *Cataclysms on the Columbia*, Timber Press, Portland, Oregon.

Allison, I.S., 1933. New version of the Spokane Flood. *Geological Society of America Bulletin* 44:675-722.

Allison, I.S., 1935. Glacial erratics in Willamette Valley. *Geological Society of America Bulletin* 46:615-632.

Allison, I.S., 1941. Flint's fill hypothesis of origin of scabland. *Journal of Geology* 49:54-73.

Alt, D.D., 1987. Multiple catastrophic drainage of glacial Lake Missoula, Montana. *Geological Society of America Centennial Field Guide -Rocky Mountain Section*, Boulder, Colorado, pp. 33-36.

Alt, D., 2001. *Glacial Lake Missoula and Its Humongous Floods*, Mountain Press Publishing, Missoula, Montana.

Alt, D. and Chambers, R.L., 1970. Repetition of the Spokane Flood. *Reports of American Quaternary Association - Abstracts of Biennial Meeting, first meeting*, p. 1

Anderson, R.Y. and Dean, W.E., 1988. Lacustrine varve formation through time. *Palaeogeography, Palaeoclimatology, Palaeoecology* 62:215-235.

Atwater, B.F., 1984. Periodic floods from glacial Lake Missoula into the Sanpoil arm of glacial Lake Columbia, northeastern Washington. *Geology* 12:464-467.

Atwater, B. F., 1986. Pleistocene glacial-lake deposits of the Sanpoil River Valley, northeastern Washington. *U.S. Geological Survey Bulletin 1661*, Washington D.C.

Atwater, B.F., 1987. Status of glacial Lake Columbia during the last floods from glacial Lake Missoula. *Quaternary Research* 27:182-201.

Austin, S.A., 1994a. A creationist view of Grand Canyon strata. In: Austin, S.A. (editor), *Grand Canyon– Monument to Catastrophism*, Institute for Creation Research, Santee, California, pp. 57-82.

Austin, S.A., 1994b. How was the Grand Canyon eroded? In: Austin, S.A. (editor), *Grand Canyon–Monument to Catastrophism*, Institute for Creation Research, Santee, California, pp. 83-110.

Austin, S.A., 1994c. Interpreting strata of Grand Canyon. In: Austin, S.A. (editor), *Grand Canyon–Monument to Catastrophism*, Institute for Creation Research, Santee, California, pp. 21-56.

Austin, S.A., 1996. Excess argon within mineral concentrates from the new dacite lava dome at Mount St. Helens volcano. *Creation Ex Nihilo Technical Journal* 10(3):335-343.

Austin, S.A., 2000. Mineral isochron method applied as a test of the assumptions of radioisotope dating. In: Vardiman, L., Snelling, A.A., and Chaffin, E.F. (editors), *Radioisotopes and the Age of the Earth:A Young-Earth Creationist Research Initiative*, Institute for Creation Research and Creation Research Society, El Cajon, California, and St. Joseph, Missouri, pp. 95-121.

Austin, S.A. and Humphreys, D.R., 1990. The sea's missing salt: a dilemma for evolutionists. In: Walsh, R.E. and Brooks, C.L. (editors), *Proceedings of the Second International Conference on Creationism*, Creation Science Fellowship, Pittsburgh, Pennsylvania, pp. 17-33.

Austin, S.A. and Snelling, A.A., 1998. Discordant potassium-argon model and isochron "ages" for Cardenas Basalt (Middle Proterozoic) and associated diabase of eastern Grand Canyon, Arizona. In: Walsh, R.E. (editor), *Proceedings of the Fourth International Conference on Creationism*, Creation Science Fellowship, Pittsburgh, Pennsylvania, pp. 35-51.

Baker, V.R., 1973. Paleohydrology and sedimentology of Lake Missoula flooding in eastern Washington. *Geological Society of America Special Paper 144*, Geological Society of America, Boulder, Colorado.

Baker, V.R., 1978a. The Spokane Flood controversy. In: Baker, V.R. and Nummedal D. (editors), *The Channeled Scabland*, NASA, Washington, D.C., pp. 3-15.

Baker, V.R., 1978b. Paleohydraulics and hydrodynamics of scabland floods. In: Baker, V.R. and Nummedal, D. (editors), *The Channeled Scabland*, NASA, Washington, D.C., pp. 59-79.

Baker, V.R., 1978c. Large-scale erosional and depositional features of the Channeled Scabland. In: Baker, V.R. and Nummedal, D. (editors), *The Channeled Scabland*, NASA, Washington, D.C., pp. 81-115.

Baker, V.R., 1978d. The Spokane Flood controversy and the Martian outflow channels. *Science* 202:1249-1256.

Baker, V.R., 1978e. Quaternary geology of the Channeled Scabland and adjacent areas. In: Baker, V.R. and Nummedal, D. (editors), *The Channeled Scabland*, NASA, Washington, D.C., pp. 17-35.

Baker, V.R., 1981. *Catastrophic Flooding: The Origin of the Channeled Scabland*. Benchmark Paper in Geology 55, Dowden, Hutchinson & Ross, Stroudsburg, Pennsylvania.

Baker, V.R., 1983. Late-Pleistocene fluvial systems. In: Wright, Jr., H.E. (editor), *Late-Quaternary Environments of the United States - Volume 1 The late Pleistocene*, University of Minnesota Press, Minneapolis, Minnesota, pp. 115-129.

Baker, V.R., 1989a. The Grand Coulee and Dry Falls. In: Breckenridge, R.M. (editor), *Glacial Lake Missoula and the Channeled Scabland*, 28th International Geological Congress Field Trip Guidebook T310, American Geophysical Union, Washington, D.C., pp. 51-55.

Baker, V.R., 1989b. Wallula Junction and Wallula Gap. In: Breckenridge, R.M. (editor), *Glacial Lake Missoula and the Channeled Scabland*, 28th International Geological Congress Field Trip Guidebook T310, American Geophysical Union, Washington, D.C., pp. 63-64.

Baker, V., 1995. Surprise endings to catastrophism and controversy on the Columbia–Joseph Thomas Pardee and the Spokane flood controversy. *GSA Today* 5(9):169-173.

Baker, V.R., 2002. The study of superfloods. *Science* 295:2379-2380.

Baker, V.R., Benito, G., and Rudoy, A.N., 1993. Paleohydrology of Late Pleistocene superflooding, Altay Mountains, Siberia. *Science* 259:348-350.

Baker, V.R. and Bunker, R.C., 1985. Cataclysmic Late Pleistocene flooding from glacial lake Missoula: a review. *Quaternary Science Reviews* 4:1-41.

Baker, V.R., Greeley, R., Komar, P.D., Swanson, D.A., and Waitt, Jr., R.B., 1987. Columbia and Snake River plains. In: Graf, W.L. (editor), *Geomorphic Systems of North America*, Geological Society of America Centennial Special Volume 2, Geological Society of America, Boulder, Colorado, p. 403-468.

Baker, V.R., Bjornstad, B.N., Busacca, A.J., Fecht, K.R., Kiver, E.P., Moody, U.L., Rigby, J.G., Stradling, D.F., and Tallman, A.M., 1991. Quaternary geology of the Columbia Plateau. In: Morrison, R.B. (editor), *Quaternary Geology of the Columbia Plateau*, The geology of North America, Vol. K-2, Quaternary Nonglacial Geology: Conterminous, U.S., Geological Society of America, Boulder, Colorado, pp. 249-250.

Barnette, D.W. and Baumgardner, J.R., 1994. Patterns of ocean circulation over the continents during Noah's Flood. In: Walsh, R.E. (editor), *Proceedings of the Third International Conference on Creationism*, Creation Science Fellowship, Pittsburgh, Pennsylvania, pp. 77-86.

Barzun, J., 1958. *Darwin, Marx, Wagner*, second edition, Doubleday & Co., Garden City, New York.

Bates, R.L. and Jackson, J.A. (editors), 1984. *Dictionary of Geological Terms*, third edition, Anchor Press/Doubleday, Garden City, New York.

Beaney, C.L., 2002. Tunnel channels in southeast Alberta, Canada: evidence for catastrophic channelized drainage. *Quaternary International* 90:67-74.

Beaney, C.L. and Shaw, J., 2000. The subglacial geomorphology of southeast Alberta: evidence for subglacial meltwater erosion. *Canadian Journal of Earth Sciences* 37:51-61.

Beget, J.E., 1986a. Comments on "Outburst floods from glacial Lake Missoula" by G.K.C. Clarke, W.H. Mathews, and R.T. Pack. *Quaternary Research* 25:136-138.

Beget, J.E., 1986b. Modeling the influence of till rheology on the flow and profile of the Lake Michigan Lobe, southern Laurentide Ice Sheet, U.S.A. *Journal of Glaciology* 32(111): 235-241.

Beget, J., 1987. Low profile of the northwest Laurentide Ice Sheet. *Arctic and Alpine Research* 19:81-88.

Begét, J.E., Keskinen, M.J., and Severin, K.P., 1997. Tephrochronologic constraints on the Late Pleistocene history of the south margin of the Cordilleran Ice Sheet, western Washington. *Quaternary Research* 47:140-146.

Benito, G., 1997. Energy expenditure and geomorphic work of the cataclysmic Missoula flooding in the Columbia River Gorge, USA. *Earth Surface Processes and Landforms* 22:457-472.

Bergman, J., 2001. Influential Darwinists supported the Nazi holocaust. *Creation Research Society Quarterly* 38(1):31-39.

Bergman, J., 2002. Darwinism as a factor in the twentieth-century totalitarianism holocausts. *Creation Research Society*

Quarterly 39(1):47-53.

Bjornstad, B.N., 1980. *Sedimentology and Depositional Environment of the Touchet Beds, Walla Walla River Basin, Washington*, Rockwell Hanford Operations Report, RHO-BWI-SA-44, Department of Energy, Richland, Washington.

Bjornstad, B.N, Fecht, K.R., and Tallman, A.M., 1991. Quaternary stratigraphy of the Pasco Basin, south-central Washington. In: Morrison, R.B. (editor), *Quaternary Geology of the Columbia Plateau*, The geology of North America, Vol. K-2, Quaternary Nonglacial Geology: Conterminous, U.S., Geological Society of America, Boulder, Colorado, pp. 228-238.

Bjornstad, B.N., Fecht, K.R., and Pluhar, C.J., 2001. Long history of pre-Wisconsin, ice age cataclysmic floods: evidence from southeastern Washington state. *Journal of Geology* 109: 695-713.

Blatt, H., Middleton, G., and Murray, R., 1972. *Origin of Sedimentary Rocks*, Prentice-Hall, Englewood Cliffs, New Jersey.

Bloom, A.L., 1971. Glacial-eustatic and isostatic controls of sea level. In: Turekian, K.K. (editor), *The Late Cenozoic Glacial Ages*, Yale University Press, New Haven, Connecticut, pp. 355-379.

Bluemle, J.P., Lord, M.L., and Hunke, N.T., 1993. Exceptionally long, narrow drumlins formed in subglacial cavities, North Dakota. *Boreas* 22:15-24.

Boyce, J.I. and Eyles, N., 1991. Drumlins carved by deforming till streams below the Laurentide ice sheet. *Geology* 19: 787-790

Boyce, J.I. and Eyles, N., 2000. Architectural element analysis applied to glacial deposits: internal geometry of a late Pleistocene till sheet, Ontario, Canada. *Geological Society of America Bulletin* 112:98-118.

Brand, L. 1997. *Faith, Reason, and Earth History*, Andrews University Press, Berrien Springs, Michigan.

Breckenridge, R.M., 1989. Lower glacial Lakes Missoula and Clark Fork ice dams. In: Beckenridge, R.M. (editor), *Glacial Lake Missoula and the Channeled Scabland*, 28th International Geological Congress Field Trip Guidebook T310, American Geophysical Union, Washington, D.C., pp. 13-21.

Breckenridge, R.M., 1993. Glacial Lake Missoula and the Spokane Floods. GeoNote 26, Idaho Geological Survey, Moscow, Idaho.

Brennand, T.A., Shaw, J., and Sharpe, D.R., 1996. Regional-scale meltwater erosion and deposition patterns, northern Quebec, Canada. *Annals of Glaciology* 22:85-92.

Bretz, J.H., 1919. The Late Pleistocene submergence in the Columbia Valley of Oregon and Washington. *Journal of Geology* 27:489-506.

Bretz, J.H., 1923a. Glacial drainage of the Columbia Plateau. *Geological Society of America Bulletin* 34:573-608.

Bretz, J.H., 1923b. The Channeled Scablands of Columbia Plateau. *Journal of Geology* 31:617-649.

Bretz, J.H., 1924a. The age of the Spokane glaciation. *American Journal of Science*, 8 (5th Ser.):336-342.

Bretz, J.H., 1924b. The Dalles type of river channel. *Journal of Geology* 32:139-149.

Bretz, J.H., 1925a. The Spokane Flood beyond the channeled scablands. I. *Journal of Geology* 33:97-115.

Bretz, J.H., 1925b. The Spokane Flood beyond the channeled scablands. II. *Journal of Geology* 33:236-259.

Bretz, J.H., 1927. The Spokane Flood: a reply. *Journal of Geology* 35: 461-468.

Bretz, J.H., 1928a. Alternative hypotheses for Channeled Scabland. I. *Journal of Geology* 36:193-223.

Bretz, J.H., 1928b. Alternative hypotheses for Channeled Scabland. II. *Journal of Geology* 36:312-341.

Bretz, J.H., 1928c. The Channeled Scabland of eastern Washington. *Geographical Review* 18:446-477.

Bretz, J.H., 1928d. Bars of Channeled Scabland. *Geological Society of America Bulletin* 39:643-702.

Bretz, J.H., 1929a. Valley deposits immediately east of the Channeled Scabland of Washington. I. *Journal of Geology* 37: 393-427.

Bretz, J.H., 1929b. Valley deposits immediately east of the Channeled Scabland of Washington. II. *Journal of Geology* 37: 505-541.

Bretz, J.H., 1930a. Valley deposits immediately west of the Channeled Scabland. *Journal of Geology* 38:385-422.

Bretz, J.H., 1930b. Lake Missoula and the Spokane Flood. *Geological Society of America Bulletin* 41:92-93.

Bretz, J.H., 1932. *The Grand Coulee*, American Geographical Society Special Publication No. 15, American Geographical Society, New York.

Bretz, J.H., 1959. *Washington's Channeled Scabland*, Washington State Division of Mines and Geology Bulletin No. 45, Olympia, Washington.

Bretz, J.H., 1969. The Lake Missoula floods and the Channeled Scabland. *Journal of Geology* 77:505-543.

Bretz, J.H., 1978. Introduction. In: Baker, V.R. and Nummedal, D. (editors), *The Channeled Scabland*, NASA, Washington, D.C., p. 1-2.

Bretz, J.H., Smith, H.T.U., and Neff, G.E., 1956. Channeled Scabland of Washington: new data and interpretations. *Geological Society of America Bulletin* 67:957-1049.

Bruce, R.H., Middleton, M.F., Holyland, P., Loewenthal, D., and Bruner, I., 1996. Modelling of petroleum formation associated with heat transfer due to hydrodynamic processes. *PESA Journal* 24:6-12.

Bunker, R.C., 1982. Evidence of multiple Late-Wisconsin floods from glacial Lake Missoula in Badger Coulee, Washington. *Quaternary Research* 18:17-31.

Carling, P.A., 1996. Morphology, sedimentology and palaeohydraulic significance of large gravel dunes, Altai Mountains, Siberia. *Sedimentology* 43:647-664.

Carpenter, K., Hirsch, K.F., and Horner, J.R. (editors), 1994. *Dinosaur Eggs and Babies*, Cambridge University Press, London.

Carson, B., 1935. Head erosion at its worst. *Soil Conservation* 1:14-15.

Carson, R.J. and Pogue, K.R., 1996. *Flood Basalts and Glacier Floods: Roadside Geology of Parts of Walla Walla, Franklin, and Columbia Counties, Washington.* Division of Geology and Earth Resources Information Circular 90, Washington State Department of Natural Resources, Olympia, Washington.

Carson, R.J., McKhann, C.F., and Pizey, M.H., 1978. The Touchet beds of the Walla Walla Valley. In: Baker, V.R. and Nummedal, D. (editors), *The Channeled Scabland*, NASA, Washington, D.C., pp. 173-177.

Chaffin, E.F., 2000. Theoretical mechanisms of accelerated radioactive decay. In: Vardiman, L., Snelling, A.A., and Chaffin, E.F. (editors), *Radioisotopes and the Age of the Earth: A Young-Earth Creationist Research Initiative*, Institute for Creation Research and Creation Research Society, El Cajon, California, and St. Joseph, Missouri, pp. 305-331.

Chaffin, E.F., 2001. A model for the variation of the Fermi constant with time. *Creation Research Society Quarterly* 38(3): 127-138.

Chambers, R.L., 1971. *Sedimentation in Glacial Lake Missoula.* M.S. thesis, University of Montana, Missoula, Montana.

Chambers, R.L., 1984. Sedimentary evidence for multiple glacial Lakes Missoula. In: McBane, J.D. and Garrison, P.B., (editors), *Northwest Montana and Adjacent Canada*, Montana Geological Society 1984 Field Conference and Symposium, Montana Geological Society, Billings, Montana, pp. 189-199.

Chambers, R.L. and Curry, R.R., 1989. Glacial Lake Missoula: sedimentary evidence for multiple drainages. In: Breckenridge, R.M. (editor), *Glacial Lake Missoula and the Channeled Scabland*, 28th International Geological Congress Field Trip Guidebook T310, American Geophysical Union, Washington, D.C., pp. 3-11.

Charlesworth, J,K., 1957. *The Quaternary Era*, Edward Arnold, London.

Clarke, G.K.C., Mathews, W.H. and Pack, R.T., 1984. Outbursts floods from glacial Lake Missoula. *Quaternary Research* 22:289-299.

Clayton, L., Teller, J.T., and Attig, J.W., 1985. Surging of the southwestern part of the Laurentide ice Sheet. *Boreas* 14: 235-241.

Clemens, J.D. and Mawer, C.K., 1992. Granitic magma transport by fracture propagation. *Tectonophysics* 204:339-360.

Coffin, H.G with Brown, R.H., 1983. *Origin by Design*, Review and Herald Publishing Association. Washington, D.C.

Colman, S.M., Keigwin, L.D., and Forester, R.M., 1994. Two episodes of meltwater influx from glacial Lake Agassiz into the Lake Michigan basin and their climatic contrasts. *Geology* 22:547-550.

Cooley, S.W., Pidduck, B.K., and Pogue, K.R., 1996. Mechanism and time of emplacement of clastic dikes in the Touchet Beds of the Walla Walla Valley, south-central Washington. *Geological Society of America Abstracts with Programs* 28:57.

Cowan, E.A., Powell, R.D., and Smith, N.D., 1988. Rainstorm-induced event sedimentation at the tidewater front of a temperate glacier. *Geology* 16:409-412.

Craig, R.G., 1987. Dynamics of a Missoula Flood. In: Mayer L. and Nash, D. (editors), *Catastrophic Flooding*, The Binghampton Symposia in Geomorphology, International Series No. 18, Allen and Unwin, Boston, pp. 305-332.

Crickmay, C.H., 1974. *The Work of the River*, Elsevier, New York.

Crickmay, C.H., 1975. The hypothesis of unequal activity. In Melhorn, W.N. and Flemel, R.C. (editors), *Theories of Landform Development*, George Allen and Unwin, London, pp. 103-109.

Dardis, G.F., McCabe, A.M., and Mitchell, W.I., 1984. Characteristics and origins of lee-side stratification sequences in Late Pleistocene drumlins, Northern Ireland. *Earth Surface Processes and Landforms* 9:409-424.

Elfström, Å., 1983. The Balkakatj boulder delta, Lapland, Northern Sweden. *Geografiska Annaler* 65A:201-225.

Elfström, Å., 1987. Large boulder deposits and catastrophic floods. *Geografiska Annaler* 69A:101-121.

Eriksson, E., 1968. Air-ocean-icecap interactions in relation to climatic fluctuations and glaciation cycles. In Mitchell, Jr., J.M. (editor), *The Causes of Climatic Change*, Meteorological Monographs 8(30), American Meteorological Society, Boston, Massachusetts, pp. 68-92.

Eyles, N., Mullins, H.T., and Hine, A.C., 1990. Thick and fast: sedimentation in a Pleistocene fiord lake of British Columbia, Canada. *Geology* 18:1153-1157.

Eyles, N., Mullins, H.T., and Hine, A.C., 1991. The seismic stratigraphy of Okanagan Lake, British Columbia:[sic] a record of rapid deglaciation in a deep 'fiord-lake' basin. *Sedimentary Geology* 73:13-41.

Faulkner, D.R., 1997. Comets and the age of the solar system. *Creation Ex Nihilo Technical Journal* 11(3):264-273.

Faure, G., 1986. *Principles of Isotope Geology*, Second Edition, John Wily & Sons, New York.

Fisher, T.G. and Shaw, J., 1992. A depositional model for Rogen moraine, with examples from the Avalon Peninsula, Newfoundland. *Canadian Journal of Earth Sciences* 29:669-686.

Flint, R.F., 1938. Origin of the Cheney-Palouse Scabland tract, Washington. *Geological Society of America Bulletin* 49: 461-524.

Froede, Jr., C.R. 1995. A proposal for a creationist geological timescale. *Creation Research Society Quarterly* 32:90-94.

Froede, Jr., C.R., 1998. *Field Studies in Catastrophic Geology*, Creation Research Society Monograph No. 7, Creation Research Society, St Joseph, Missouri.

Fryxell, R., 1962. A radiocarbon limiting date for scabland flooding. *Northwest Science* 36(4):113-119.

Fulthorpe, C.S. and Austin, Jr., J.A., 1998. Anatomy of rapid margin progradation: three-dimensional geometries of Miocene clinoforms, New Jersey margin. *American Association of Petroleum Geologists Bulletin* 82(2):251-273.

Gansser, A., 1964. *Geology of the Himalayas*, Interscience Publishers, New York.

Gardner, J.V., Field, M.E., and Twichell, D.C. (editors), 1996. *Geology of the United States' Seafloor–The View from GLORIA*, Cambridge University Press, Cambridge, UK.

Garner, P., 1996. Where is the Flood/post-Flood boundary? Implications of dinosaur nests in the Mesozoic. *Creation Ex Nihilo Technical Journal* 10(1):101-106.

Garton, M., 1996. The pattern of fossil tracks in the geological record. *Creation Ex Nihilo Technical Journal* 10(1):82-100.

Gilbert, R., 1990. Evidence for the subglacial meltwater origin and late Quaternary lacustrine environment of Bateau Channel, eastern lake Ontario. *Canadian Journal of Earth Sciences* 27:939-945.

Gilbert, R., 2000. The Devil Lake pothole (Ontario): evidence of subglacial fluvial processes. *Géographie Physique et Quaternaire* 54:245-250.

Gilbert, R. and Shaw, J., 1994. Inferred subglacial meltwater origin of lakes on the southern border of the Canadian Shield. *Canadian Journal of Earth Sciences* 31:1630-1637.

Gish, D.T., 1995. *Evolution: The Fossils Still Say No!* Institute for Creation Research, El Cajon, California.

Glenn, J.L., 1965. *Late Quaternary Sedimentation and Geologic History of the North Willamette Valley, Oregon*, PhD thesis, Oregon State University, Corvallis, Oregon.

Gould, S.J., 1987. *Time's Arrow, Time's Cycle*, Harvard University Press, Cambridge, Massachusetts.

Graham, R.W. and Lundelius, Jr., E.L., 1984. Coevolutionary disequilibrium and Pleistocene extinctions. In: Martin, P.S. and Klein, R.G. (editors), *Quaternary Extinctions: A Prehistoric Revolution*, University of Arizona Press, Tuscon, Arizona, pp. 223-249.

Gupta, H.K., Radhakrishna, I., Chadha, R.K., Kümpel, H.-J., and Grecksch, G., 2000. Pore pressure studies initiated in area of reservoir-induced earthquakes in India. *EOS* 81(14): 145,151.

Hall, K., 1982. Rapid deglaciation as an initiator of volcanic activity: an hypothesis. *Earth Surface Processes and Landforms* 7:45-51.

Haller, Jr., J.S., 1971. *Outcasts from Evolution: Scientific Attitudes of Racial inferiority, 1859-1900*, McGraw-Hill, New York.

Ham, K., Wieland, C., and Batten, D., 1999. *One Blood: The Biblical Answer to Racism*, Master Books, Green Forest, Arkansas.

Hanson, L.G., 1970. *The Origin and Development of Moses Coulee and Other Scabland Features on the Waterville Plateau, Washington*, PhD thesis, University of Washington, Seattle, Washington.

Hindmarsh, R.C.A., 1998. Drumlinization and drumlin-forming instabilities: viscous till mechanisms. *Journal of Glaciology* 44(147):293-314.

Hill, D.P. *et al.*, 1993. Seismicity remotely triggered by the magnitude 7.3 Landers, California, earthquake. *Science* 260: 1617-1623.

Hill, D.P., Pollitz, F., and Newhall, C., 2002. Earthquake–volcano interactions. *Physics Today* 55(11):41-47.

Hill, G.R., 1972. Some aspects of coal research. *Chemical*

Technology 292-297.

Himmelfarb, G., 1962. *Darwin and the Darwinian Revolution*, W.W. Norton & Company, New York.

Hobbs, W.H., 1943. Discovery in eastern Washington of a new lobe of the Pleistocene continental glacier. *Science* 98: 227-230.

Hodge, E.T., 1934. Origin of the Washington scablands. *Northwest Science* 8:4-11.

Horner, J.R., 1982. Evidence for colonial nesting and 'site fidelity' among ornithischian dinosaurs. *Nature* 297:675-676.

Horner, J.R. and Currie, P.J., 1994. Embryonic and neonatal morphology and ontogeny of a new species of *Hypacrosaurus* (Ornithischia, Lambeosauridae) from Montana and Alberta. In: Carpenter, K., Hirsch, K.F., and Horner, J.R. (editors), *Dinosaur Eggs and Babies*, Cambridge University Press, London, pp. 312-336.

Horner, J.R. and Dobb, E., 1997. *Dinosaur Lives–Unearthing an Evolutionary Saga*, Harper Collins, New York.

Horner, J.R. and Weishampel, D.B., 1996. A comparative embryological study of two ornithischian dinosaurs. *Nature* 383:103.

Hoyle, F., 1981. *Ice the Ultimate Human Catastrophe.* Continuum, New York.

Humphreys, D.R., 2000. Accelerated nuclear decay: a viable hypothesis? In: Vardiman, L., Snelling, A.A., and Chaffin, E.F. (editors), *Radioisotopes and the Age of the Earth: A Young-Earth Creationist Research Initiative*, Institute for Creation Research and Creation Research Society, El Cajon, California, and St. Joseph, Missouri, pp. 333-379.

Hutton, D.H.W., Dempster, T.J., Brown, P.E., and Becker, S.D., 1990. A new mechanism of granite emplacement: intrusion in active extensional shear zones. *Nature* 343:452-455.

Jaffee, M.A. and Spencer, P.W., 2000. Multiple floods many times over: the record of glacial outburst floods in southeastern Washington. *Geological Society of America Abstracts with Programs* 32:A21.

Jarrett, R.D. and Malde, H.E., 1987. Paleodischarge of the late Pleistocene Bonneville Flood, Snake River, Idaho, computed from new evidence. *Geological Society of America Bulletin* 99: 127-134.

Jenkins, O.P., 1925. Clastic dikes of eastern Washington and their geologic significance. *American Journal of Science*, 10(5th series):234-246.

Johnson, P.E., 1991. *Darwin on Trial*, Regnery Gateway, Washington, D.C.

Karrow, P.F., Seymour, K.L., Miller, B.B., and Mirecki, J.E., 1997. Pre-Late Wisconsinan Pleistocene biota from southeastern Michigan, U.S.A. *Palaeogeography, Palaeoclimatology, Palaeoecology* 133:81-101.

Kehew, A.E., 1982. Catastrophic flood hypothesis for the origin of the Souris spillway, Saskatchewan and North Dakota. *Geological Society of America Bulletin* 93:1051-1058.

Kehew, A.E. and Lord, M.L., 1986. Origin and large-scale erosional features of glacial-lake spillways in the northern Great Plains. *Geological Society of America Bulletin* 97:162-177.

Kehew, A.E. and Lord, M.L., 1987. Glacial-lake outbursts along the mid-continent margins of the Laurentide ice-sheet. In: Mayer, L. and Nash, D. (editors), *Catastrophic Flooding*, Allen and Unwin, Boston, Massachusetts, pp. 95-120.

Kehew, A.E. and Teller, J.T., 1994. History of late glacial runoff along the southwestern margin of the Laurentide Ice Sheet. *Quaternary Science Reviews* 13:859-877.

Kennett, J., 1982, *Marine Geology*, Prentice-Hall, Englewood cliffs, New Jersey.

King, L.C., 1967. *The Morphology of the Earth–A Study and Synthesis of World Scenery*, Hafner Publishing Company, New York.

King, L.C. 1982. *The Natal Monocline– Explaining the Origin and Scenery of Natal, South Africa*, second revised edition, University of Natal Press, Pietermaritzburg, South Africa.

King, L.C., 1983. *Wandering Continents and Spreading Sea Floors on an Expanding Earth*, John Wiley and Sons, New York.

Kiver, E.P., Stradling, D.F. and Baker, V.R., 1989. The Spokane Valley and northern Columbia Plateau. In: Breckenridge, R.M. (editor), *Glacial Lake Missoula and the Channeled Scabland*, 28th International Geological Congress Field Trip Guidebook T310, American Geophysical Union, Washington, D.C., pp. 23-35.

Klevberg, P. and Oard, M.J., 1998. Paleohydrology of the Cypress Hills Formation and Flaxville gravel. In: Walsh, R.E. (editor), *Proceedings of the Fourth International Conference in Creationism*, Creation Science Fellowship, Pittsburgh, Pennsylvania, pp. 361-378.

Komatsu, G., Miyamoto, H., Ito, K., Tosaka, H., and Tokunaga, T., 2000. Comment - the Channeled Scabland: back to Bretz? comment and reply. *Geology* 28:573-574.

Kor, P.S.G. and Cowell, D.W., 1998. Evidence for catastrophic subglacial meltwater sheetflood events on the Bruce Peninsula, Ontario. *Canadian Journal of Earth Sciences* 35:1180-1202.

Lambert, A. and Hsü, K.J., 1979. Non-annual cycles of varve-like sedimentation in Walensee, Switzerland. *Sedimentology* 26:453-461.

Leckie, D.A. and Cheel, R.J., 1989. The Cypress Hills Formation (Upper Eocene to Miocene): a semiarid braidplain deposit resulting from intrusive uplift. *Canadian Journal of Earth Sciences*, 26:1918-1931.

Lidmar-Bergström, K., Olsson, S., and Olvmo, M., 1997. Palaeosurfaces and associated saprolites in southern Sweden. In: Widdowson, M. (editor), *Palaeosurfaces: Recognition, Reconstruction and Palaeoenvironmental Interpretation*, Geological Society of London Special Publication No. 120, pp. 95-124.

Linde, A.T. and Sacks, S., 1998. Triggering of volcanic eruptions. *Nature* 395:888-890.

Lister, J.C., 1981. *The Sedimentology of Camas Prairie Basin and Its Significance to the Lake Missoula Floods*, M.S. thesis, University of Montana, Missoula, Montana.

Lockley, M., 1991. *Tracking Dinosaurs–A New Look at an Ancient World*, Cambridge University Press, London.

Lockley, M.G., 1994. Dinosaur ontogeny and population structure: interpretations and speculations based on fossil footprints. In: Carpenter, K., Hirsch, K.F., and Horner, J.R. (editors), *Dinosaur Eggs and Babies*, Cambridge University Press, London, pp. 347-365.

Lockley, M. and Hunt, A.P., 1995. *Dinosaur Tracks and Other Fossil Footprints of the Western United States*, Columbia University Press, New York.

Lubenow, M.L., 1992. *Bones of Contention–A Creationist Assessment of Human Fossils*, Baker Book House, Grand Rapids, Michigan.

Lucchitta, I., 1990. History of the Grand Canyon and of the Colorado River in Arizona. In: Beus, S.S. and Morales, M. (editors), *Grand Canyon Geology*, Oxford University Press, New York, pp. 311-332.

Lupher, R.L., 1944. Clastic dikes of the Columbia Basin region, Washington and Idaho. *Geological Society of America Bulletin* 55:1431-1462.

Mackiewicz, N.E., *et al.*, 1984. Interlaminated ice-proximal glaciomarine sediments in Muir Inlet, Alaska. *Marine Geology* 53:113-147.

Martin, P.S. and Klein, R.G. (editors), 1984. *Quaternary Extinctions: A Prehistoric Revolution*, University of Arizona Press, Tuscon, Arizona.

Mathews, W.H., 1974. Surface profiles of the Laurentide ice sheet in its marginal areas. *Journal of Glaciology* 13(67):37-43.

McCabe, A.M. and Dardis, G.F., 1989. A geological view of drumlins in Ireland. *Quaternary Science Reviews* 8:169-177.

McCabe, A.M., Knight, J., and McCarron, S.G., 1999. Ice-flow states and glacial bedforms in north central Ireland: a record of rapid environmental change during the last glacial termination. *Journal of the Geological Society, London* 156:63-72.

McDonald, E.V. and Busacca, A.J., 1988. Record of pre-late Wisconsin giant floods in the Channeled Scabland interpreted from loess deposits. *Geology* 16:728-731.

McGuire, W.J., Howarth, R.J., Firth, C.R., Solow, A.R., Pullen, A.D., Saunders, S.J., Stewart, I.S., and Vita-Finzi, C., 1997. Correlation between rate of sea-level change and frequency of explosive volcanism in the Mediterranean. *Nature* 389: 473-476.

McKee, B., 1972. *Cascadia: The Geological Evolution of the Pacific Northwest*, McGraw-Hill, New York.

McKnight, E.T., 1927. The Spokane Flood: a discussion. *Journal of Geology* 35:453-460.

Moody, U.L., 1987. *Late Quaternary Stratigraphy of the Channeled Scabland and Adjacent Areas*, PhD thesis, University of Idaho, Moscow, Idaho.

Morris, H.M., 1989. *The Long War against God: The History and Impact of the Creation/Evolution Conflict*, Baker Book House, Grand Rapids, Michigan.

Morris, J.D., 1994. *The Young Earth*, Master Books, Green Forest, Arkansas.

Morris, J.D., 2002. A canyon in six days! *Creation* 24(4):54-55.

Mullineaux, D.R., 1986. Summary of pre-1980 tephra-fall deposits erupted from Mount St. Helens, Washington state, USA. *Bulletin of Volcanology* 48:17-26.

Mullineaux, D.R., Wilcox, R.E., Ebaugh, W.F., Fryxell, R., and Rubin M., 1978. Age of the last major scabland flood of the Columbia Plateau in Eastern Washington. *Quaternary Research* 10:171-180.

Munro-Stasiuk, M.J., 2000. Rhythmic till sedimentation: evidence for repeated hydraulic lifting of a stagnant ice mass. *Journal of Sedimentary Research* 70(1):94-106.

Munro-Stasiuk, M.J. and Shaw, J., 2002. The Blackspring Ridge flute field, south-central Alberta, Canada: evidence for subglacial sheetflow erosion. *Quaternary International* 90: 75-86.

Mustoe, G.E., 2001. *Skolithos* in a quartzite cobble from Lopez Island–are Western Washington's oldest fossils Canadian emigrants? *Washington Geology* 29(3/4):17-19.

Nakada, M. and Yokose, H., 1992. Ice age as a trigger of active Quaternary volcanism and tectonism. *Tectonophysics* 212: 321-329.

Nott, J. and Roberts, R.G., 1996. Time and process rates over the past 100 Ma: a case for dramatically increased landscape denudation rates during the late Quaternary in northern Australia. *Geology*, 24:883-887.

Nott, J., Young, R.W., and McDougall, I., 1996. Wearing down, wearing back, and gorge extension in the long-term denudation of a highland mass; quantitative evidence from the Shoalhaven catchment, south-east Australia. *Journal of Geology*, 104:224-332.

Oard, M.J., 1990. *An Ice Age Caused by the Genesis Flood*, Institute for Creation Research, El Cajon, California.

Oard, M.J., 1992a. Varves - the first "absolute" chronology: part I– historical development and the question of annual deposition. *Creation Research Society Quarterly* 29(2):72-80.

Oard, M.J., 1992b. Varves - the first "absolute" chronology: part II–varve correlation and the post-glacial time scale. *Creation Research Society Quarterly* 29(3):120-125.

Oard, M.J., 1995. Polar dinosaurs and the Genesis Flood. *Creation Research Society Quarterly* 32:47-56.

Oard, M.J., 1996a. Where is the Flood/post-Flood boundary in the rock record? *Creation Ex Nihilo Technical Journal* 10(2): 258-278.

Oard, M.J., 1996b. Are those 'old' landforms in Australia really old? *Creation Ex Nihilo Technical Journal* 10(2):174-175.

Oard, M.J., 1996c. K-Ar dating results in major landform surprises. *Creation Ex Nihilo Technical Journal* 10(3):298-299.

Oard, M.J., 1997a. *Ancient Ice Ages or Gigantic Submarine Landslides?* Creation Research Society Monograph No. 6, Creation Research Society, St Joseph, Missouri.

Oard, M.J., 1997b. New dating method calculates unreasonably low rates of granite erosion in Australia. *Creation Ex Nihilo Technical Journal* 11(2):128-130.

Oard, M.J., 1997c. The extinction of the dinosaurs. *Creation Ex Nihilo Technical Journal* 11(2):137-154.

Oard, M.J., 1997d. *The Weather Book*, Master Books, Green Forest, Arkansas.

Oard, M.J., 1998a. Australian landforms: consistent with a young earth. *Creation Ex Nihilo Technical Journal* 12(3):253-254.

Oard, M.J, 1998b. Dinosaurs in the Flood: a response. *Creation Ex Nihilo Technical Journal* 12(1):69-86.

Oard, M.J., 1998c Were the Colorado valleys cut during post-Flood or Flood times? *Creation Research Society Quarterly* 35(2):104-107.

Oard, M.J., 2000a. Only one Lake Missoula flood. *Creation Ex Nihilo Technical Journal* 14(2):14-17.

Oard, M.J., 2000b. Antiquity of landforms: objective evidence that dating methods are wrong. *Creation Ex Nihilo Technical Journal* 14(1):35-39.

Oard, M.J., 2001a. Vertical tectonics and the drainage of Floodwater: a model for the middle and late Diluvian period–part I. *Creation Research Society Quarterly* 38(1):3-17.

Oard, M.J., 2001b. Vertical tectonics and the drainage of Floodwater: a model for the middle and late Diluvian period–part II. *Creation Research Society Quarterly* 38(2):79-95.

Oard, M.J., 2003. In the footsteps of giants. *Creation* **25**(2): 10-12.

Oard, M.J. and Klevberg, P., 1998. A diluvial interpretation of the Cypress Hills Formation, Flaxville gravel, and related deposits. In: Walsh, R.E. (editor), *Proceedings of the Fourth International Conference in Creationism*, Creation Science Fellowship, Pittsburgh, Pennsylvania, pp. 421-436.

Oberlander, T., 1965. *The Zagros Streams–A New Interpretation of Transverse Drainage in an Orogenic Zone*, Syracuse Geographical Series No. 1., Syracuse, New York.

Oberlander, T., 1985. Origin of drainage transverse to structure in orogens. In: Morisawa, M. and Hack, J.T. (editors), *Tectonic Geomorphology*, Allen and Unwin, Boston, Massachusetts, pp. 155-182.

O'Cofaigh, C.O., 1996. Tunnel valley genesis. *Progress in Physical Geography* 20(1):1-19.

O'Connor, J.E., 1993. *Hydrology, Hydraulics, and Geomorphology of the Bonneville Flood*, Geological Society of America Special Paper 274, Geological Society of America, Boulder, Colorado.

O'Connor, J.E. and Baker, V.R., 1992. Magnitudes and implications of peak discharges from glacial Lake Missoula. *Geological Society of America Bulletin*, 104:267-279.

O'Connor, J.E. and Waitt, R.B., 1995a. Beyond the Channeled Scabland–a field trip to Missoula flood features in the Columbia, Yakima, and Walla Walla Valleys of Washington and Oregon–part 1. *Oregon Geology*, 57(3):51-60.

O'Connor, J.E. and Waitt, R.B., 1995b. Beyond the Channeled Scabland–a field trip to Missoula flood features in the Columbia, Yakima, and Walla Walla Valleys of Washington and Oregon–part 3:field trip, days two and three. *Oregon Geology*, 57(5):99-115.

Ollier, C.D., Gaunt, G.F.M., and Jurkowski, I., 1988. The Kimberly Plateau, Western Australia: a Precambrian erosion surface. *Zeitschrift für Geomorpholgie N.F.* 32:239-246.

Ollier, C. and Pain, C., 2000. *The Origin of Mountains*, Routledge, New York.

Olson, E.C., 1969. Introduction to J. Harlen Bretz's paper

on "the Lake Missoula floods and the Channeled Scabland." *Journal of Geology* 77:503-504.

Ovenshine, A.T., 1970. Observations of iceberg rafting in Glacier Bay, Alaska, and the identification of ancient ice-rafted deposits. *Geological Society of America Bulletin* 81: 891-894.

Paltridge, G.W. and Platt, G.M.R., 1976. *Radiation Processes in Meteorology and Climatology*, Elsevier, New York.

Pardee, J.T., 1910. The glacial Lake Missoula. *Journal of Geology* 18:376-386.

Pardee, J.T., 1940. Ripple marks (?) in glacial Lake Missoula, Montana. *Geological Society of America Bulletin* 51:2028-2029.

Pardee, J.T., 1942. Unusual currents in glacial Lake Missoula, Montana. *Geological Society of America Bulletin* 53:1569-1600.

Parfit, M., 1995. The floods that carved the west. *Smithsonian*, 26(1):48-59.

Paterson, W.S.B., 1981. *The Physics of Glaciers*, 2nd edition, Pergamon Press, New York.

Patton, P.C. and Baker, V.R., 1978a. Origin of the Cheney-Palouse scabland tract. In: Baker, V.R. and Nummedal, D. (editors), *The Channeled Scabland*, NASA, Washington, D.C., pp.117-130.

Patton, P.C. and Baker, V.R., 1978b. New evidence for pre-Wisconsin flooding in the Channeled Scabland of eastern Washington. *Geology* 6:567-571.

Patton, P.C., Baker, V.R., and Kochel, R.C., 1979. Slack-water deposits: a geomorphic technique for the interpretation of fluvial paleohydrology. In: Rhodes, D.D. and Williams, G.P. (editors), *Adjustments of the Fluvial System*, Kendal/Hunt Publishing Co., Dubuque, Iowa, pp. 225-253.

Pendick, D., 1996. The dust ages. *Earth* 5(3):22-23,66-67.

Perloff, J., 1999. *Tornado in a Junkyard: The Relentless Myth of Darwinism*, Refuge Books, Arlington, Massachusetts.

Pettijohn, F.J., 1975. *Sedimentary Rocks*, third edition, Harper and Row, New York.

Pickard, J., 1984. Comments on "Wastage of the Klutlan ice-cored moraines, Yukon Territory, Canada" by Driscoll (1980). *Quaternary Research* 22(2):259.

Pickrill, R.A. and Irwin, J., 1983. Sedimentation in a deep glacier-fed lake-Lake Tekapo, New Zealand. *Sedimentology* 30:63-75.

Pinna, G., 1985. *The Illustrated Encyclopedia of Fossils*, Facts on File, New York.

Piotrowski, J.A., 1987. Genesis of the Woodstock drumlin field, southern Ontario, Canada. *Boreas* 16:249-265.

Plummer, C.C. and McGeary, D., 1996. *Physical Geology*, seventh edition, Wm. C. Brown Publishers, Dubuque, Iowa.

Porter, D.A. and Guccione, M.J., 1994. Deglacial flood origin of the Charleston alluvial fan, Lower Mississippi alluvial valley. *Quaternary Research* 41:278-284.

Power, J.A., Moran, S.C., McNutt, S.R., Stihler, S.D., and Sanchez, J.J., 2001. Seismic response of the Katmai volcanoes to the 6 December 1999 magnitude 7.0 Karluk Lake earthquake, Alaska. *Bulletin of the Seismological Society of America* 91(1):57-63.

Quigley, R.M., 1983. Glaciolacustrine and glaciomarine clay deposition: a North American perspective. In: Eyles, N. (editor), *Glacial Geology–An Introduction for Engineers and Earth Scientists*, Pergamon Press, New York, pp. 140-167.

Rains, B., Shaw, J., Skoye, R., Sjogren, D., and Kvill, D., 1993. Late Wisconsin subglacial megaflood paths in Alberta. *Geology* 21:323-326.

Rains, R.B., Shaw, J., Sjogren, D.B., Munro-Stasiuk, M.J., Skoye, K.R., Young, R.R., and Thompson, R.T., 2002. Subglacial tunnel channels, Porcupine Hills, southwest Alberta, Canada. *Quaternary International* 90:57-65.

Ramm, B. 1954. *The Christian View of Science and Scripture*, William B. Eerdmans Publishing, Grand Rapids, Michigan.

Rampino, M.R. and Self, S., 1993. Climate-volcanism feedback and the Toba eruption of ~74,000 years ago. *Quaternary Research* 40:269-280.

Rampton, V.N., 2000. Large-scale effects of subglacial meltwater flow in the southern Slave Province, Northwest Territories, Canada. *Canadian Journal of Earth Sciences* 37:81-93.

Raup, D.M. and Stanley, S.M., 1978. *Principles of Paleontology*, second edition, W.H. Freeman and Company, San Francisco, California.

Reed, J.K. (editor), 2000. *Plate Tectonics: A Different View*. Creation Research Society Monograph No. 10. Creation Research Society, St. Joseph, Missouri.

Rice, Jr., J.W. and Edgett, K.S., 1997. Catastrophic flood sediments in Chryse Basin, Mars, and Quincy Basin, Washington: application of sandar facies model. *Journal of Geophysical Research*, 102 (E2):4185-4200.

Richmond, G.M., Fryxell, R., Neff, G.E., and Weis, P.L., 1965. The Cordilleran ice sheet of the Northern Rocky Mountains and related Quaternary history of the Columbia Plateau. In: Wright, Jr., H.E. and Frey, D.G. (editors), *The Quaternary of the Untied States*, Princeton University Press, Princeton, New Jersey, pp. 230-242.

Rigby, J.G., 1982. *The sedimentology, mineralogy, and depositional environment of a sequence of Quaternary catastrophic flood-derived lacustrine turbidites near Spokane, Washington*, M.S. thesis, University of Idaho, Moscow, Idaho..

Roth, A.A., 1998. *Origins–Linking Science and Scripture*, Review and Herald Publishing, Hagerstown, Maryland.

Rudoy, A.N., 2002. Glacier-dammed lakes and geological work of glacial superfloods in the Late Pleistocene, Southern Siberia, Altai Mountains. *Quaternary International* 87:119-140.

Rudwick, M.J.S., 1990. Introduction to Charles Lyell's *Principles of Geology*, University of Chicago Press, Chicago, Illinois, pp. vii-lviii.

Rugg, S.H. and Austin, S.A.,1998. Evidences for rapid formation and failure of Pleistocene "lava dams" of the western Grand Canyon, Arizona. In: Walsh, R.E. (editor), *Proceedings of the Fourth International Conference on Creationism*, Creation Science Fellowship, Pittsburgh, Pennsylvania, pp. 475-486.

Russell, A.J. and Knudsen, O., 1999. An ice-contact rhythmite (turbidite) succession deposited during the November 1996 catastrophic outburst flood (jökulhlaup), Skeioarárjökull, Iceland. *Sedimentary Geology* 127:1-10.

Schermerhorn, L.J.G., 1974. Late Precambrian mixtites: glacial and/or nonglacial? *American Journal of Science* 274:673-824.

Schumm, S., 1963. Disparity between present rates of denudation and orogeny. *U.S. Geological Survey Professional Paper 454*, U.S. Geological Survey, Washington, D.C.

Schumm, S. and Ethridge, F.G., 1994. Origin, evolution and morphology of fluvial valleys. In: Dalrymple, R.W., Boyd, R., and Zaitlin, B.A. (editors), *Incised-Valley Systems: Origins and Sedimentary Sequences*, SEPM Special Publication No. 51, Tulsa, Oklahoma, pp. 11-27.

Sharpe, D.R. and Shaw, J., 1989. Erosion of bedrock by subglacial meltwater, Cantley, Quebec. *Geological Society of America Bulletin* 101:1011-1020.

Shaver, M., 1998. Flood geology sheds light on Unaweep Canyon mystery. *Creation Research Society Quarterly* 34(4): 218-224.

Shaw, J., 1983. Drumlin formation related to inverted meltwater erosional marks. *Journal of Glaciology* 29(103):461-479.

Shaw, J., 1988a. Subglacial erosional marks, Wilton Creek, Ontario. *Canadian Journal of Earth Sciences* 25:1256-1267.

Shaw, J., 1988b. Pyroclasts: nothing new, nothing new. *Geoscience Canada* 15(4):291-292.

Shaw, J., 1989. Drumlins, subglacial meltwater floods, and ocean responses. *Geology* 17:853-856.

Shaw, J., 1996. A meltwater model for Laurentide subglacial landscapes. In: McCann, S.B. (editor), *Geomorphology Sans Frontières*, John Wiley & Sons, New York, pp. 181-236.

Shaw, J., 2002. The meltwater hypothesis for subglacial bedforms. *Quaternary International* 90:5-22.

Shaw, J., Faragini, D.M., Kvill, D.R., and Rains, R.B., 2000. The Athabasca fluting field, Alberta, Canada: implications for the formation of large-scale fluting (erosional lineations). *Quaternary Science Reviews* 19:959-980.

Shaw, J. and Gilbert, R., 1990. Evidence for large-scale subglacial meltwater flood events in southern Ontario and northern New York State. *Geology* 18:1169-1172.

Shaw, J. and Kvill, D., 1984. A glaciofluvial origin for drumlins of the Livingstone Lake area, Saskatchewan. *Canadian Journal of Earth Sciences* 21:1442-1459.

Shaw, J., Kvill, D., and Rains, B., 1989. Drumlins and catastrophic subglacial floods. *Sedimentary Geology* 62:177-202.

Shaw, J., Munro-Stasiuk, M., Sawyer, B., Beanery, C., Lesemann, J.-E., Musacchio, A., Rains, B., and Young, R.R., 1999. The Channeled Scabland: back to Bretz? *Geology* 27: 605-608.

Shaw, J., Munro-Stasiuk, M., Sawyer, B., Beanery, C., Lesemann, J.-E., Musacchio, A., Rains, B., and Young, R.R., 2000. The Channeled Scabland: back to Bretz?–reply. *Geology* 28:576.

Shaw, J., Rains, B., Eyton, R., and Weissling, L, 1996. Laurentide subglacial outburst floods: landform evidence form digital elevations models. *Canadian Journal of Earth Sciences* 33:1154-1168.

Shaw, J. and Sharpe, D.R., 1987. Drumlin formation by subglacial meltwater erosion. *Canadian Journal of Earth Sciences* 24:2316-2322.

Shilts, W.W., 1980. Flow patterns in the central North American Ice Sheet. *Nature* 286:213-218.

Shilts, W.W., Cunningham, C.J., and Kaszycki, C.A., 1979. Keewatin Ice Sheet–re-evaluation of the traditional concept of the Laurentide Ice Sheet. *Geology* 7:537-541.

Shoemaker, E.M., 1992. Water sheet outburst floods from the Laurentide Ice Sheet. *Canadian Journal of Earth Sciences* 29: 1250-1264.

Shoemaker, E.M., 1995. On the meltwater genesis of drumlins. *Boreas* 24(3):3-10.

Shoemaker, E.M., 1999. Subglacial water-sheet floods, drumlins and ice-sheet lobes. *Journal of Glaciology* 45:201-213.

Sibrava, V., Bowen, D.Q., and Richmond, G.M. (eds), 1986. *Quaternary Glaciations in the Northern Hemisphere*, Pergamon Press, New York.

Sieja, D.M., 1959. *Clay Mineralogy of Glacial Lake Missoula Varves, Missoula County, Montana*, M.S. thesis, University of Montana, Missoula, Montana.

Sigvaldason, G.E., Annertz, K., and Nilsson, M., 1992. Effect of glacial loading/deloading on volcanism: postglacial volcanic production rate of the Dyngjufjöll area, central Iceland. *Bulletin of Volcanology* 54:385-392.

Simpson, S., 1998. Big shocks push volcanoes over the edge. *Science News* 154:279.

Sissons, J.B., 1979. Catastrophic lake drainage in Glen Spean and the Great Glen, Scotland. *Journal of the Geological Society, London* 136:215-224.

Sjogren, D.B. and Rains, R.B., 1995. Glaciofluvial erosional morphology and sediments of the Coronation– Spondin Scabland, east-central Alberta. *Canadian Journal of Earth Sciences* 32:565-578.

Smith, D.G., and Fisher, T.G., 1993. Glacial lake Agassiz: the northwestern outlet and paleoflood. *Geology* 21:9-12.

Smith, G.A., 1993. Missoula flood dynamics and magnitudes inferred from sedimentology of slack-water deposits on the Columbia Plateau, Washington. *Geological Society of America Bulletin* 105:77-100.

Smith, G.I., and Street-Perrott, F.A., 1983. Pluvial lakes of the Western United States. In: Wright, Jr., H.E. (editor), *Late-Quaternary Environments of the United States*, volume 1, University of Minnesota Press, Minneapolis, Minnesota, pp. 190-212.

Smith, N.D., Phillips, A.C. and Powell, R.D., 1990. Tidal drawdown: a mechanism for producing cyclic sediment laminations in glaciomarine deltas. *Geology* 18:10-13.

Snelling, A.A., 1994. U-Th-Pb "dating": an example of false "isochrons," In: Walsh, R.E. (editor), *Proceedings of the Third International Conference on Creationism*, Creation Science Fellowship, Pittsburgh, Pennsylvania, pp. 497-504.

Snelling, A.A., 1995. The failure of U-Th-Pb "dating" at Koongarra, Australia. *Creation Ex Nihilo Technical Journal* 9(1):71-92.

Snelling, A.A., 1998. The cause of anomalous potassium-argon "ages" for recent andesite flows at Mt Ngauruhoe, New Zealand, and the implications for potassium-argon "dating." In: Walsh, R.E. (editor), *Proceedings of the Fourth International Conference on Creationism*, Creation Science Fellowship, Pittsburgh, Pennsylvania, pp. 503-525.

Snelling, A.A., 2000. Geochemical processes in the mantle and crust. In: Vardiman, L., Snelling, A.A., and Chaffin, E.F. (editiors), *Radioisotopes and the Age of the Earth: A Young-Earth Creationist Research Initiative*, Institute for Creation Research and Creation Research Society, El Cajon, California, and St. Joseph, Missouri, pp. 123-304.

Spencer, P.K., 1989. A small mammal fauna from the Touchet beds of Walla Walla County, Washington: support for the multiple-flood hypothesis. *Northwest Science* 63(4):167-174.

Stevenson, A.J., Scholl, D.W., and Vallier, T.L., 1983. Tectonic and geologic implications of the Zodiac fan, Aleutian abyssal plain, northeast Pacific. *Geological Society of America Bulletin* 94:259-273.

Strahler, A.N., 1987. *Science and Earth History–The Evolution/ Creation Controversy*, Prometheus Books, Buffalo, New York.

Sugden, D.E. and John, B.S., 1976. *Glaciers and Landscape: A Geomorphological Approach*, Edward Arnold, London.

Summerfield, M.A., 1991. *Global Geomorphology*, Longman Scientific and Technical and John Wiley and Sons, New York.

Taylor, C., 1998. Did mountains really rise according to Psalm 104:8? *Creation Ex Nihilo Technical Journal* 12(3):312-313.

Taylor, C., 1999. More on mountains– Charles Taylor replies. *Creation Ex Nihilo Technical Journal* 13(1):70-71.

Teller, J.T., 1987. Proglacial lakes and the southern margin of the Laurentide Ice Sheet. In Ruddiman, W.F. and Wright, Jr., H.E. (editors), *The Geology of North America, Vol. K-3, North America and Adjacent Oceans during the Last Deglaciation*, Geological Society of America, Boulder, Colorado, pp. 39-69.

Teller, J.T., 1995. History and drainage of large ice-dammed lakes along the Laurentide Ice Sheet. *Quaternary International* 28:83-92.

Teller, J.T. and Thorliefson, L.H., 1987. Catastrophic flooding into the Great lakes from Lake Agassiz. In: Mayer, L. and Nash, D. (editors), *Catastrophic Flooding*, Allen and Unwin, Boston, Massachusetts, pp. 121-138.

Thornbury, W.D., 1965. *Regional Geomorphology of the United States*, John Wiley and Sons, New York, NY.

Thorson, R.M., 1996. Earthquake recurrence and glacial loading in western Washington. *Geological Society of America Bulletin* 108:1182-1191.

Tinkler, K.J. and Stenson, R.E., 1992. Sculpted bedrock forms along the Niagara escarpment, Niagara Peninsula, Ontario. *Géographie Physique et Quaternaire* 46(2):195-207.

Tolan, T.L., Reidel, S.P., Beeson, M.H., Anderson, J.L., Fecht, K.R., and Swanson, D.A., 1989. Revisions to the estimates of the areal extent and volume of the Columbia River Basalt Group. In: Reidel, S.P. and Hooper, P.R. (editors), *Volcanism and Tectonism in the Columbia River Flood-Basalt Province*,

Geological Society of America Special Paper 239, Geological Society of America, Boulder, Colorado, pp. 1-20.

Twidale, C.R., 1998. Antiquity of landforms: an 'extremely unlikely' concept vindicated. *Australian Journal of Earth Sciences*, 45:657-668.

Vallier, T., 1998. *Islands and Rapids–A Geologic Story of Hells Canyon*, Confluence Press, Lewiston, Idaho.

Vanderburgh, S. and Roberts, M.C., 1996. Depositional system and seismic stratigraphy of a Quaternary basin: North Okanagan Valley, British Columbia. *Canadian Journal of Earth Sciences* 33:917-927.

Vardiman, L., Snelling, A.A., and Chaffin, E.F. (editors), 2000. *Radioisotopes and the Age of the Earth: A Young-Earth Creationist Research Initiative*, Institute for Creation Research and Creation Research Society, El Cajon, California, and St. Joseph, Missouri.

Varricchio, D.J., Jackson, F., Borkowski, J.J., and Horner, J.R., 1997. Nest and egg clutches of the dinosaur *Troödon formosus* and the evolution of avian reproductive traits. *Nature* 385: 247-250.

Ver Steeg, K., 1930. Wind gaps and water gaps of the northern Appalachians, their characteristics and significance. *Annals of the New York Academy of Sciences* 32:87-220.

Vonhof J.A., 1965. The Cypress Hills Formation and its reworked deposits in southwestern Saskatchewan. In: Zell, R.L. (editor), *15th Annual Field Conference Guidebook, Part I Technical Papers, Part I Cypress Hills Plateau Alberta and Saskatchewan*, Alberta Society of Petroleum Geologists, Calgary, Alberta, pp 142-161.

Waitt, Jr., R.B., 1980. About forty last-glacial Lake Missoula jökulhlaups through southern Washington. *Journal of Geology* 88:653-679.

Waitt, Jr., R.B., 1984. Periodic jökulhlaups from Pleistocene glacial Lake Missoula–new evidence from varved sediments in northern Idaho and Washington. *Quaternary Research* 22: 46-58.

Waitt, Jr., R.B., 1985. Case for periodic, colossal jökulhlaups from Pleistocene glacial Lake Missoula. *Geological Society of America Bulletin* 96:1271-1286.

Waitt, R.B., 1994. Scores of gigantic, successively smaller Lake Missoula floods through Channeled Scabland and Columbia Valley. In: Swanson, D.A. and Haugerud, R.A. (editors), *Geologic Field Trips in the Pacific Northwest*, volume 1, Department of Geological Sciences, University of Washington, pp. k1-k88.

Waitt, R.B., and Atwater, B.F., 1989. Stratigraphic and geomorphic evidence for dozens of last-glacial floods. In: Breckenridge, R.M. (editor), *Glacial Lake Missoula and the Channeled Scabland*, 28th International Geological Congress Field Trip Guidebook T310, American Geophysical Union, Washington, D.C., pp. 37-50.

Waitt, Jr., R.B. and Thorson, R.M., 1983. The Cordilleran Ice Sheet in Washington, Idaho, and Montana. In: Wright, Jr., H.E. (editor), *Late-Quaternary Environments of the United States–Volume 1 The Late Pleistocene*, University of Minnesota Press, Minneapolis, Minnesota, pp. 53-70.

Walker, E.H., 1967. Varved lake beds in northern Idaho and northeastern Washington. *U.S. Geological Survey Professional Paper 574-B*, U.S. Government Printing Office, Washington, D.C., p. B83-B87.

Walker, T., 1994. A Biblical geological model. In: Walsh, R.E. (editor), *Proceedings of the Third International Conference on Creationism*, Creation Science Fellowship, Pittsburgh, Pennsylvania, pp. 581-592.

Walker, T., 1996a. The basement rocks of the Brisbane area, Australia: where do they fit in the creation model. *Creation Ex Nihilo Technical Journal* 10(2):241-257.

Walker, T., 1996b. The Great Artesian Basin, Australia. *Creation Ex Nihilo Technical Journal* 10(3):379-390.

Ward, P.D., 1997. *The Call of Distant Mammoths–Why the Ice Age Mammoths Disappeared*, Springer-Verlag, New York.

Watson, T., 1997. What causes ice ages? *U.S. News & World Report*, 123(7):58-60.

Weber, W.M., 1972. *Correlation of Pleistocene Glaciation in the Bitterroot Range, Montana, with Fluctuations of Glacial Lake Missoula*, Bureau of Mines and Geology Memoir 42, Montana Bureau of Mines and Geology, Butte, Montana.

Westgate, J.A. and Naeser, N.D., 1985. Dating methods of Pleistocene deposits and their problems: V. tephrochronology and fission-track dating. In: Rutter, N.W. (editor), *Dating Methods of Pleistocene Deposits and Their Problems*, Geological Association of Canada, Toronto, Ontario, pp. 31-38.

Whitcomb, Jr., J.C. and Morris, H.M., 1961. *The Genesis Flood*, Baker Book House, Grand Rapids, Michigan.

Williams, E.L. 1998. Rapid canyon formation: the Black Canyon of the Gunnison River, Colorado. *Creation Research Society Quarterly* 35:148-155.

Williams, E.L., 1999. Unaweep Canyon–another visit. *Creation Research Society Quarterly* 36:155-156.

Williams, E.L., Meyer, J.R. and Wolfrom, G.W., 1991. Erosion of Grand Canyon of the Colorado River: part I– review of antecedent river hypothesis and the postulation of large quantities of rapidly flowing water as the primary agent of erosion. *Creation Research Society Quarterly* 28:92-98.

Williams, E.L., Meyer, J.R. and Wolfrom, G.W., 1992a. Erosion of the Grand Canyon of the Colorado River: part II– review of river capture, piping and ancestral river hypotheses and the possible formation of vast lakes. *Creation Research Society Quarterly* 28:138-145.

Williams, E.L., Meyer, J.R. and Wolfrom, G.W., 1992b. Erosion of the Grand Canyon of the Colorado River: part III– review of the possible formation of basins and lakes on Colorado Plateau and different climatic conditions in the past. *Creation Research Society Quarterly* 29:18-24.

Williams, E.L., Chaffin, E.F., Goette, R.M. and Meyer, J.R., 1994. Pine Creek Gorge, the Grand Canyon of Pennsylvania: an introductory creationist study. *Creation Research Society Quarterly* 31:44-59.

Williams, L.D., 1979. An energy balance model of potential glacierization of northern Canada. *Arctic and Alpine Research* 11:445-456.

Wood, A.E., 1947. Multiple banding of sediments deposited during a single season. *American Journal of Science* 245:304-312.

Woodmorappe, J., 1999a. *Studies in Flood Geology–A Compilation of Research Studies Supporting Creation and the Flood*, Institute for Creation Research. El Cajon, California.

Woodmorappe, J., 1999b. *The Mythology of Modern Dating Methods*, Institute for Creation Research, El Cajon, California.

Woodmorappe, J., 1999c. Radiometric geochronology reappraised. In: Woodmorappe, J., *Studies in Flood Geology–A Compilation of Research Studies Supporting Creation and the Flood*, Institute for Creation Research, El Cajon, California, pp. 145-175.

Wonderly, D.E., 1987. *Neglect of Geological Data: Sedimentary Strata Compared with Young-Earth Creationist Writings*, Interdisciplinary Biblical Research Institute, Hatfield, Pennsylvania.

Wright, G.F., 1911. *The Ice Age in North America*, Bibliotheca Sacra Co., Oberlin, Ohio.

Young, D.A., 1982. *Christianity & the Age of the Earth*, Zondervan, Grand Rapids, Michigan.

Other Creation Research Society Books

Scientific Studies in Special Creation
edited by Walter E. Lammerts, Ph.D.

Speak to the Earth
edited by George F. Howe, Ph.D.

The Argument: Creation Vs. Evolutionism
by Wilbert H. Rusch, Sr., M.S., LL.D.

Thermodynamics and the Development of Order
edited by Emmett L. Williams, Ph.D.

Did Charles Darwin Become a Christian?
by Wilbert H. Rusch, Sr., M.S., LL.D. and John W. Klotz, Ph.D.

Creationist Research
(1964-1980), by Duane T. Gish, Ph.D.

Vestigial Organs Are Fully Functional
by Jerry Bergman, Ph.D. and George F. Howe, PhD.

Origins: What Is at Stake?
by Wilbert H. Rusch, Sr., M.S., LL.D.

Astronomy and Creation
by Donald B. DeYoung, Ph.D.

Ancient Ice Ages or Gigantic Submarine Landslides?
by Michael J. Oard, M.S.

Physical Science and Creation
by Donald B. DeYoung, Ph.D.

Design and Origins in Astronomy
edited by George Mulfinger, M.S.

Design and Origins in Astronomy, Volume 2
edited by Donald B. DeYoung, Ph.D. and Emmett L. Williams, Ph.D.

Field Studies in Catastrophic Geology
by Carl R. Froede, Jr., P.G.

The North American Midcontinent Rift System
by John K. Reed, Ph.D.

The Human Body: An Intelligent Design
by Alan L. Gillen, Ed.D., Frank J. Sherwin III, M.A., and Allan Knowles, M.S.

Plate Tectonics: A Different View
edited by John K. Reed, Ph.D.

Natural History in the Christian Worldview
by John K. Reed, Ph.D.

Science and Creation
by Wayne Frair, Ph.D.

Biology and Creation
by Wayne Frair, Ph.D.

For more information about the Creation Research Society, a book catalog, or a subscription to the *Creation Research Society Quarterly* contact:

Creation Research Society
6801 N. Highway 89
Chino Valley, AZ 86323-9186
www.creationresearch.org
Email: contact@creationresearch.org